Explorations in Learning and the Brain

Ton de Jong · Tamara van Gog · Kathleen Jenks ·
Sarah Manlove · Janet van Hell · Jelle Jolles ·
Jeroen van Merriënboer · Theo van Leeuwen ·
Annemarie Boschloo

Explorations in Learning and the Brain

On the Potential of Cognitive Neuroscience for Educational Science

Springer

Ton de Jong
Faculty of Behavioral Sciences
University of Twente
7500 AE Enschede
Netherlands
A.J.M.deJong@utwente.nl

Tamara van Gog
Centre for Learning Sciences and Technologies
Open University of the Netherlands
P.O. Box 2960
6401 DL Heerlen
Netherlands
tamara.vangog@ou.nl

Kathleen Jenks
Behavioural Science Institute
Radboud University Nijmegen
Montessorilaan 3
6525 HR Nijmegen
Netherlands
k.jenks@pwo.ru.nl

Sarah Manlove
University of Twente
7500 AE Enschede
Netherlands
manlovesa@gmail.com

Janet van Hell
Behavioral Science Institute
Radboud University Nijmegen
Montessorilaan 3
6525 HR Nijmegen
Netherlands
j.vanhell@pwo.ru.nl

Jelle Jolles
Faculty of Psychology and Education\
VU University Amsterdam
van der Boechorststraat 1
1081 BT Amsterdam
Netherlands
j.jolles@psy.vu.nl

Jeroen van Merriënboer
Open University of the Netherlands
6401 DL Heerlen
Netherlands
jeroen.vanmerrienboer@ou.nl

Theo van Leeuwen
Faculty of Behavioral Sciences
University of Twente
7500 AE Enschede
Netherlands
t.h.vanleeuwen@utwente.nl

Annemarie Boschloo
Faculty of Psychology and Education
VU University Amsterdam
van der Boechorststraat 1
1081 BT Amsterdam
Netherlands
am.boschloo@psy.vu.nl

ISBN 978-0-387-89511-6 e-ISBN 978-0-387-89512-3
DOI 10.1007/978-0-387-89512-3

Library of Congress Control Number: 2009924163

Printed on acid-free paper

springer.com

Preface

This volume presents a short review study of the potential relationships between cognitive neuroscience and educational science. Conducted by order of the Dutch Programme Council for Educational Research of the Netherlands Organization for Scientific Research (NWO; cf. the American NSF), the review aims to identify: (1) how educational principles, mechanisms, and theories could be extended or refined based on findings from cognitive neuroscience, and (2) which neuroscience principles, mechanisms, or theories may have implications for educational research and could lead to new interdisciplinary research ventures.

The contents should be seen as the outcome of the 'Explorations in Learning and the Brain' project. In this project, we started with a 'quick scan' of the literature that formed the input for an expert workshop that was held in Amsterdam on March 10–11, 2008. This expert workshop identified additional relevant themes and issues that helped us to update the 'quick scan' into this final document. In this way the input from the participants of the expert workshop (listed in Appendix A) has greatly influenced the present text. We are therefore grateful to the participants for their scholarly and enthusiastic contributions. The content of the current volume, however, is the full responsibility of the authors.

This project was of a modest size and as such this resulting volume is not intended to present a comprehensive view of the field. Instead, it tries to name a number of interesting research alleys on the crossroad of educational science and cognitive neuroscience and we hope that in this respect it helps to build a research agenda.

The Netherlands Ton de Jong

Contents

Chapter 1
Introduction

The past decade has seen efforts on the part of research, education and policy communities to create a dialogue about the potential relationship between cognitive neuroscience and both the science and practice of education. Notable examples include the publications from the Centre for Educational Research and Innovation (CERI) of the OECD. Their 2002 report on learning sciences and the brain was recently followed by a report entitled 'Understanding the brain: The birth of a learning science' (2007). This book evaluated state of the art knowledge and insights from the cognitive sciences and neurosciences which are pertinent to education. It gives an agenda for the future development of this field and encourages collaboration between the learning sciences, brain research, and policy organizations. Likewise, the report 'Brain Lessons' (Jolles et al., 2006) and its earlier version 'Learning to know the Brain' (Jolles et al., 2005), published under the auspices of the Netherlands Organisation for Scientific Research, stated that the time is ripe for an active exchange between scientists from neuroscience, cognitive science, educational science and the practice of education. Berninger and Richard's (2002) book on Brain Literacy reaches out to educators and psychologists about what we know of the brain and how it might be relevant to teaching and learning. Another example is a recent report of the German Ministry of Education that, after reviewing relevant neuroscientific research, concluded with ten research questions that link neuroscience and educational science (Stern, Grabner, & Schumacher, 2006). The status of this dialogue is also underpinned by the launch of a new journal (Mind, Brain, and Education) in 2007. In the Netherlands, the Netherlands Organisation for Scientific Research in collaboration with the Ministry of Education, Culture and Science installed the 'Brain and Learning committee' in 2003, which organized an invitational conference on this multi-dimensional research domain in 2004, which led to the above mentioned report. A further Dutch initiative is currently being undertaken by the 'Study Centre for Technology Trends' (van Keulen & Rispens, in press). Review articles such as Byrnes and Fox (1998), Goswami (2004) and Goswami (2006), Posner and Rothbart (2005), Katzir and Paré-Blagoev (2006), and Varma, McCandliss, and Schwartz (2008) furthered this dialogue by asking critical questions about the educational implications of cognitive neuroscience research. New initiatives include bibliometric analyses to explore

T. de Jong et al., *Explorations in Learning and the Brain*,

DOI 10.1007/978-0-387-89512-3_1,

whether there already are overlaps between the two fields in the research literature (Merkx & van Koten, in press).

This volume presents a review study, conducted by order of the Dutch Programme Council for Educational Research of the Netherlands Organisation for Scientific Research, that also seeks to contribute to the afore mentioned dialogue. It provides a review of the present state of potential relationships that exist between cognitive neurosciences and educational sciences.

The present volume takes a somewhat different stance than most of the studies mentioned in the first paragraph in the sense that it does not take developments within neuroscience as the starting point but rather sets off from major questions that are dominant in educational research, notably instructional systems design and related fields within the educational sciences. It's goals are to identify interesting interfaces between neuroscientific and educational research, as well as to inform the program council on potentially interesting additions to educational research programs of the Netherlands Organisation for Scientific Research and viable interdisciplinary ventures. By drawing on empirical findings from both disciplines, the following general questions will be addressed:

1. Which principles, mechanisms and theories studied in educational research could be further extended or refined based on findings from cognitive neuroscience?
2. Which principles, mechanisms and theories studied in cognitive neuroscience may have implications for educational research?
3. What are these implications and which (interdisciplinary or transdisciplinary) research questions can be drawn from them?
4. What form could an interdisciplinary or transdisciplinary research program take based on research questions generated from the above questions?

The approach taken is that of a literature review which does not pretend to provide a complete coverage of the pertinent domain and does not give an in-depth evaluation of all the relevant issues. Rather, it highlights some of the most important trends that can be observed in the recent literature.

The literature review we made to answer the questions mentioned above developed as follows. First, a short list of educational topics was developed to begin an initial scan of the literature. This scan resulted in the creation of a more constrained list of relevant articles and journals. This list was reviewed by cognitive neuroscience and learning science experts prior to the actual literature review to ensure accuracy. A review was then conducted to gather relevant empirical findings from both fields. In view of the purpose of the report, there was a focus upon research papers published in recent years, notably those which are relevant for the major topics investigated in present-day educational research. This resulted in an initial report, which we labelled a 'quick scan', and that formed the input for an expert workshop held in Amsterdam, March 10–11, 2008.[1] The workshop helped to

[1] Participants in the Amsterdam workshop are listed in Appendix A.

review the previously identified themes as well as to identify new themes and issues that are now included in the current, final, volume. We should emphasize, though, that given the size of the current project the overview of the literature as presented is still limited. As such, the current, final, volume is organised around these common topics and issues:

1. Learning principles, including effects of multimodal processing, learning from multiple representations, cognitive load, the role of insight in problem solving, implicit and explicit learning, self-directed learning including the role of regulative skills and metacognition, and the use of observation and /or imitation for learning to perform practical and cognitive tasks
2. The role of affective processes in learning
3. Learning specific domains, such as (second) language learning and mathematics,
4. Learning problems, including dyslexia and dyscalculia

In addition, the current text also describes two issues (plasticity and maturation) from cognitive neuroscience that are relevant for education and that more or less traverse through the topics mentioned above (see e.g., Bach-y-Rita, Danilov, Tyler, & Grimm, 2005; Merzenich et al., 1996; Taub, 2004). As mentioned, this list is not at all meant to be exhaustive but rather provides a focus upon areas commonly addressed in the literature and which have potential for fruitful collaboration between the fields of educational research and cognitive neuroscience.

The following chapters detail the above topic definitions, describe relevant cognitive neuroscientific research and discuss how these findings may contribute to learning science theory or research. Where applicable, future directions are presented to advance potential research avenues between cognitive neuroscience and learning science communities.

Chapter 2
Learning Principles

In past decades, educational research has put major efforts into the development and evaluation of 'modern' learning environments which are characterized by an emphasis on (structured) self-directed learning, the acquisition of insightful, conceptual, knowledge, and collaborative learning. They also embed the content in a (multimodal and multi-representational) realistic context (Mayer, 2001). The current chapter focuses on the representation of learning materials in relation to the functioning of the brain, the role of cognitive load in learning, the role of implicit learning, characteristics of insightful knowledge, higher order skills associated with (structured) self-regulated learning and learning in social situations through observation and imitation. Though the latter may also be considered an aspect of collaborative learning, to our knowledge, collaborative learning is an aspect from contemporary educational theories that has not yet been addressed by neuroscientific research.

2.1 Multimodal Processing

2.1.1 Education

Dual coding theory (Paivio, 1979, 1986) states that recall is enhanced by presenting information in both visual and verbal form. The theory assumes that there are two cognitive subsystems, one specialized for the representation and processing of nonverbal information and the other specialized for dealing with language. Baddeley's model of working memory states that there is a central executive and two separate 'slave' systems for dealing with auditory and visual information, the phonological loop and the visuo-spatial sketchpad, respectively (Baddeley & Hitch, 1974). Later, another component was added: the episodic buffer (Baddeley, 2000). Although there has been criticisms of Paivio's theory (see e.g., Fliessbach, Weis, Klaver, Elger, & Weber, 2006; van Hell & de Groot, 1998), dual coding theory often forms the basis of educational design. Inspired by Paivio's and Baddeley's work, research on multimedia learning has tested the assumption that spreading information over auditory and visual modalities (pictures/animations and spoken text) leads to lower cognitive

T. de Jong et al., *Explorations in Learning and the Brain*,
DOI 10.1007/978-0-387-89512-3_2, © Springer Science+Business Media, LLC 2009

load on working memory and better learning outcomes than presenting information in a single modality (pictures/animations with written text). These results were often found (at least under restricted time conditions) and have come to be known as the 'modality effect' (see Low & Sweller, 2005).

2.1.2 Cognitive Neuroscience

In cognitive psychology, a distinction is made between verbal and non-verbal working memory, and within both types, between auditory and visual working memory. However, as is also the case in Paivio's theory, not all authors define their terms very clearly and sometimes grey areas remain (e.g., would visual stimuli that can be named be classified as verbal or non-verbal?). In recent years there has been a tremendous amount of research investigating how aspects of working memory, verbal learning, and how the use of strategies and/or the organization of memory performance are related to brain function through the use of functional brain imaging (fMRI or PET).[1] The following account provides three examples of cognitive neuroscience research pertinent to educational science and practice.

Beauchamp, Lee, Argall, and Martin (2004) found an enhanced activation of the posterior superior temporal sulcus and middle temporal gyrus (pSTS/MTG) when auditory and visual object features (of man-made objects (tools) and animals) were presented together, as compared to presentation in a single modality. Crottaz-Herbette, Anagnoson, and Menon (2004) investigated similarities and differences between visual verbal working memory and auditory verbal working memory. Their findings suggest that although similar regions are involved in both auditory and visual verbal working memory, there are modality differences in the way in which neural signals are generated, processed, and routed. Another study that is interesting in this respect comes from Kirchhoff and Buckner (2006). In an attempt to explain differences in memory abilities between individuals, they used fMRI to investigate the effects of the use of different encoding strategies on memory performance (in their study: retrieval of object associations). They showed that individuals' use of verbal elaboration and visual inspection strategies independently correlated with memory performance as operationalised by retrieval of object associations and that these strategies engage distinct brain regions that may separately influence memory performance.

---

[1] In this volume, basically three techniques for measuring brain activity are mentioned. *fMRI* (Functional magnetic resonance imaging) is a neuroimaging technique that registers changes in blood flow and blood oxygenation in the brain (haemodynamic response) related to neural activity. fMRI's are acquired in an fMRI scanner. *PET* (Positron Emission Tomography) images of the brain are also taken in a scanner and also images blood flow in the brain. In PET, a radioactive isotope must be injected into the blood stream. *EEG* (Electroencephalography) measures electrical activity produced by the brain via electrodes that are placed on the scalp. EEG measurement has a higher temporal resolution than fMRI and PET, and, in contrast to fMRI and PET techniques, EEG allows for data acquisition in natural settings.

2.1.3 Future Directions

The findings by Beauchamp et al. (2004) were based on features that are different in modality but belong to the same object (e.g., animal, tool) and were relatively simple, so the question remains whether this finding would hold, for example, for a stimulus consisting of spoken text and picture about a certain topic. Investigating implications for the redundancy effect (e.g., presenting the same text in written and spoken form should hamper processing as compared to using one representation) from a neural perspective would also be interesting, as the findings by Crottaz-Herbette et al. (2004) suggest that the same brain regions are activated in response to stimuli in auditory and visual verbal working memory but different processes occur.

2.2 Learning from Multiple Representations

2.2.1 Education

A representation is something that stands for something else (Palmer, 1978) and nowadays many such representations, usually conveying the same information, are combined to form multiple representations, for example in textbooks, where text and illustrations (photographs, and, or line drawings) try to convey a message to students or in multimedia environments where (interactive) videos, text, diagrams and other representations are combined.

In an overview on learning with such multiple representations de Jong et al. (1998) mention three reasons for introducing more than one type of representation in one learning environment. These reasons concern aspects of specificity, expertise, and sequence. According to de Jong et al. (1998) information that is specific for a certain topic should be displayed in a format that is best suited for that topic, hence in a specific representation. Given the vast variety of information to be conveyed in a complete set of learning materials, this would require several types of representations. de Jong et al.'s second reason for the use of multiple representation concerns expertise because, according to the authors, expertise is quite often seen as the possession and coordinated use of multiple representations of the same domain. A third reason for the use of multiple representations is based upon the assumption that a specified sequence of learning materials is beneficial for the learning process (de Jong et al., 1998). When learning with such multiple representations, learners are confronted with several tasks. They have to learn to understand the particulars of each separate representation, they have to understand the relation between the representation and the domain it is representing, and they have to understand the relation between separate representations (cf. Ainsworth, Bibby, & Wood, 1997). The beneficial effects of multiple representations depend on various factors, including the specific type of multiple representation employed (i.e., concurrent presentation or transitional (dynamic linking) presentations), type of domain to which the learning material belongs, the type of test used to assess the effect, subject variables, and aspects related to instructional help (see e.g., Seufert, 2003; van der Meij & de Jong, 2006).

2.2.2 Cognitive Neuroscience

The debate in cognitive neuroscience concerning the question as to whether there are separate representations associated with different input modalities (e.g., Paivio, 1991) or whether inputs from different modalities combine into a common (set of) representations (e.g., Rapp, Hillis, & Caramazza, 1993) is still unsettled (see also Section 2.1 on multimodal processing). Multimodal processing is not necessarily implicated in multiple representations, because the latter processing is usually restricted to the visual domain (albeit that visual information, particularly words, may to some extent illicit processing related to the auditory domain, e.g., inner speech). Sometimes, however, multimodality is not implemented with respect to the different primary sensory domains (e.g., visual or auditory), but as instances of different representations of the same cognitive concept within a single sensory modality. For example, in studying semantics and its neurophysiological representation in the brain, words and pictures presented in the visual domain have been shown to partially share a neuronal substrate. The anterior part of the fusiform gyrus was implicated in the representation of conceptual knowledge, irrespective of the modality of the visual input (visual word or picture), meaning that there is a single semantic representation, which also commonly recruited the left parahippocampal and perirhinal cortex and the left inferior frontal gyrus, but word-specific activations were found in the anterior temporal cortex and picture-specific activations in the occipitotemporal cortex (Bright, Moss, & Tyler, 2004; Vandenberghe, Price, Wise, Josephs, & Frackowiak, 1996).

Similarly, with respect to the concept of numeracy, the single cognitive concept of the number 2, can be activated by an auditory representation (the spoken word 'two') or visual representations of the word 'two' or the Arabic digit '2' or maybe by presenting two instances of some arbitrary visual or auditory stimulus. Domain specific brain correlates have been found in the horizontal segment of the intraparietal sulcus, a bilateral region in the posterior superior parietal lobule involved in visuospatial and attentional processes. In addition, activation was observed in the left angular gyrus (also know as the visual word form area) and left-hemispheric perisylvian areas, which are not specific to the number domain but relate to aspects of language, including verbal coding (Dehaene, Piazza, Pinel, & Cohen, 2003) (see also Chapter 5).

These two examples of multiple representations are relevant to the question of the effect of multiple representations in the learning environment, in the sense that they illustrate single concepts (semantics, number), comparable to the single conceptual message or information that has to be communicated in the multi-representational learning environment. However, there are also (fundamental) differences, because in the former examples stimuli are usually statically and successively presented, whereas in the multi representational learning situations, stimulation can be either static or dynamic and multiple representations of the information are usually simultaneously presented.

2.2.3 Future Directions

There are several relevant issues in cognitive neuroscience that may help to clarify the neurophysiological underpinnings of learning with multiple representations and may help to explain how these representations are cognitively processed and how they lead to deeper understanding.

One of these topics relates to selective attention. The different representations (assuming that they are simultaneously present) cannot be simultaneously processed in a conscious way. It would therefore be advantageous to know what determines the order in which they are processed and to find out whether this order determines the quality of processing. One relevant question in this respect concerns, for example, whether (initial or subsequent) selection of different representations relates to the amount of redundancy of the information that is conveyed by them ('complementary functions', Ainsworth, 1999).

Another relevant aspect related to the processing of multiple representations concerns memory load and attentional resources (see e.g., Bunge, Klingberg, Jacobsen, & Gabrieli, 2000). Neurophysiological research could clarify how simultaneously presented multiple representations share such resources, how they impose memory load, and how this relates to task demands. Issues from the dual task literature may be relevant here, although in such paradigms subjects explicitly have to ignore one (secondary) task in favour of another (primary) task, while in the case of multiple representations, subjects are free to divide their (limited capacity) attentional resources.

Research could also aim toward unravelling the contribution of separate representations underlying the simultaneous representation. Does processing a graph differ from that of a concrete instance or animation of the same topic (when controlled for differences in, for example, dynamics and physical visual features)? Insight into this matter could be obtained by studying synchronized networks in the brain, as manifested in patterns of cortical coherence (see e.g., Tallon-Baudry & Bertrand, 1999). Relevant questions include: Are these patterns stable over time or do their dynamics reveal aspects of learning? Do separate coherence patterns, for separate representations, sum up to the coherence pattern of the multiple representations?

In addition, future research should also aim at disentangling maintenance of longer term goal directed processing from that of separate (transient) processing of the multiple representations that have to be inspected to achieve a goal (cf. Dosenbach et al., 2007).

Further, simultaneously presenting different representations of the same (or highly similar) information may lead to redundancy and possibly to competition with respect to access to the neural substrate for elaborate processing, especially when there is an overlap in the brain areas involved, as has been shown in multimodal processing (see Section 2.1). Future research should shed light on this issue.

2.3 Cognitive Load

2.3.1 Education

Cognitive load theory (CLT, Sweller, 1988; Sweller, van Merriënboer, & Paas, 1998; van Merriënboer & Sweller, 2005) proposes that in order to be effective, learning materials should be designed in a way that takes human cognitive architecture into account. The model of cognitive architecture used in CLT consists of a working memory that is limited in capacity and time when it comes to holding or processing novel information (see Cowan, 2001; Miller, 1956), and a long-term memory with virtually unlimited capacity. Working memory limitations regarding novel information are a bottleneck when it comes to learning. Only 7 +/– 2 information elements can be held in working memory, and the number decreases (Cowan, 2001) when information not only has to be remembered (e.g., word lists), but also processed (i.e., when elements inter-relate and have to be combined, as in solving a math problem). However, information that has already been stored in long-term memory (in the form of cognitive schemata) can be handled in working memory as a single information element. Therefore, having prior knowledge (or expertise) of a certain task lowers the cognitive load imposed by that task, leaving more capacity available for other processes (e.g., deeper elaboration). Moreover, when a task or aspects of a task are repeatedly practiced (i.e., with increasing expertise), cognitive schemata become automated, and no longer require controlled processing (Schneider & Shiffrin, 1977; Shiffrin & Schneider, 1977), which further frees up working memory resources. In sum, prior domain knowledge or expertise leads to more efficient processing.

2.3.2 Cognitive Neuroscience

Neuroscience research into the mechanisms underlying cognitive load has been done in the past decade using PET, fMRI, and EEG. Using PET, Jonides et al. (1997) found that increases in task difficulty on a verbal task were associated with decreases in performance and increases in activation patterns in verbal working memory regions. Using fMRI, Jansma, Ramsey, Slagter, and Kahn (2001) showed that automatic processing that occurs due to repeated practice of a task is visible on a behavioural level in faster, less variable, and more accurate responses. At a neural level, such automatic processing results in a decrease in activation in the regions related to working memory. It should be noted, however, that these authors found no evidence for a shift of foci within or across regions of the brain. In addition to domain expertise, a higher level of intelligence also seems to be associated with higher efficiency of processing (neural efficiency). Using EEG, Grabner, Neubauer, and Stern (2006) studied the effects of chess players' intelligence and expertise on tasks related to mental speed, memory and reasoning (half the tasks were chess related, the other half were not). They concluded that intelligence and expertise influenced the efficiency of brain processing independently of each other. Participants with higher (figural) intelligence displayed a lower amount of cortical activation (interpreted as an indication of higher efficiency) than less intelligent participants, and (figural) intelligence did not lose its impact on neural efficiency when expertise is involved. Interestingly, expertise did have effects (more focused activation patterns) on the speed and reasoning tasks, but not on the memory tasks. The authors speculated that this might be due to the activation of a larger knowledge base, the use of more deliberate strategies, or both. They also indicate that it remains unclear whether this can be regarded as an indicator of neural efficiency.

2.3.3 Future Directions

The finding that higher intelligence is associated with higher neural efficiency is very interesting, but raises a causality question. Grabner et al.'s (2006) findings on the memory task show that neuroscientific methods might have the same drawback as the cognitive load measures do when used in educational research: the fact that certain regions are activated to a certain extent (or that a certain amount of load is imposed) does not always reveal which cognitive processes are occurring (or impose the load), and uncovering these processes is crucial for understanding learning or performance outcomes. Thus, an important route for future research is to deepen our understanding of brain activation patterns in relation to particular cognitive and learning processes, for instance, through the combined use of measurements from neuroscience and quantitative and qualitative (but potentially subjective) process measures such as eye movement data or thinking aloud protocols and retrospective reports (cf. van Gog, Paas, & van Merriënboer, 2005; van Gog, Paas, van Merriënboer, & Witte, 2005). New experimental paradigms to combine neurocognitive measures and measures of learning processes and educational performance should be developed.

2.4 Insightful Problem Solving

2.4.1 Education

In education there is a shift of attention from problems that can be solved in an algorithmic way to problems that require insight and conceptual knowledge (Bransford, Brown, & Cocking, 1999). According to Bowden, Jung-Beeman, Fleck, and Kounios (2005), 'insight solutions differ from non-insight solutions in a number of ways: (i) solvers experience these solutions as sudden and obviously correct (the Aha!), (ii) prior to producing an insight solution, solvers sometimes come to an impasse, no longer progressing towards a solution, and (iii) solvers usually cannot report the processing that enables them to overcome an impasse and reach a solution' (pp. 322–323). According to Ohlsson (1992) 'insight occurs in the context of an impasse, which is unmerited in the sense that the thinker is, in fact, competent to solve the problem' (p. 4). In other words, if you do not have the prerequisite knowledge you can never have insight. However, insights can also be wrong. Moreover, insight also has a subjective component: one can have insight experiences (Aha Erlebnis) on non-insight, (incremental) problems (e.g., some algebra problems).

Insight occurs usually due to a shift in problem representation, or re-representation. Often problems are not adequately represented, because not all relevant information is available. Another important consideration is the extent to which the representation enables inferences to be drawn: it may be unknown what relevant inferences are, or the representation may not enable any inferences at all. A well-known insight problem is Maier's (1931) two strings problem. His experiment also showed that small hints (the experimenter 'accidentally' brushing against the strings) can lead to a re-representation, without the learners even realizing they had got a hint.

Metcalfe (1986) and Metcalfe and Wiebe (1987) have shown that for non-insight problems, students were able to give warmth ratings that increased every few seconds, indicating they were coming closer to a solution. For insight problems however, these ratings did not increase until just before the solution was found, suggesting that insight occurs suddenly. This was further corroborated by Jausovec and Bakracevic (1995) who demonstrated that heart rate during problem solving is also dependent on the kind of problem solved. Incremental problems are accompanied by a steadily increasing heart rate, while insight problems can be recognized by a steady heart rate that suddenly increases at the end (supposedly when the insight occurs). This seems strong evidence for a difference between finding solutions in incremental non-insight problems and genuine insight problems.

2.4.2 Cognitive Neuroscience

Jung-Beeman et al. (2004) conducted a study on neural activity during insight and non-insight problem solving. They note that some questions about insight persist: whether unconscious processing precedes reinterpretation and solution, whether distinct cognitive and neural mechanisms beyond a common problem-solving network are involved in insight, and whether the apparent suddenness of insight solutions reflects truly sudden changes in cognitive processing and neural activity.

Jung-Beeman et al. (2004) hypothesize that the anterior superior temporal gyrus of the right hemisphere (RH), which is important for recognizing distant semantic relations, might play an important role in insight on verbal remote association problems. This hypothesis is based on previous studies by two of the authors (Bowden & Beeman, 1998). In that study, they found that when people were presented with potential solution words for the association task (actual solution and unrelated words), while working on a verbal problem they had yet to solve, the actual solution words were read faster than the unrelated words, and that this effect was larger when words were presented in the left visual hemifield, meaning they were projected into the RH. This RH advantage occurred only when solvers experienced insight. Trials consisted of verbal association problems in which three words were presented and the task was to find a single word that could combine with each of the three words to form new words (e.g., pine, crab, sauce → apple). Participants were asked whether they experienced insight or not, and differences in processing of insight and non-insight solutions were investigated. In their first experiment, fMRI was used. Participants solved 59% of the problems. Of the solved problems, they indicated (by a bimanual button press and subsequent verbalization of the solution word) solving 56% with insight, 41% without insight (and 2% other). As predicted, insight solutions were associated with greater neural activity in the RH anterior superior temporal gyrus (aSTG) than non-insight solutions. Although insight solutions may sometimes produce a strong emotional response, this is not likely to be due to the insight itself, as the area also showed increased activation when participants first encountered each problem. No insight effects occurred in the temporal cortex of the LH, and involvement of the RH did not appear to be due to greater difficulty in producing insight solutions given that solution times did not differ for insight and non-insight solutions.

In a second experiment, Jung-Beeman et al. (2004) investigated whether insight really occurs suddenly as studies by Metcalfe (1986), for example, suggest. They used EEG because of its greater temporal resolution. They expected to see a sudden increase in high-frequency gamma band oscillations in electrodes over the RH aSTG just before insight. In this experiment, 46% of problems were solved correctly, 56% of those reportedly with insight. A burst of high-frequency gamma band activity was associated with correct insight solutions, but not with non-insight solutions, approximately 0.3 seconds before the button was pressed to indicate the solution. Again, there was no difference between insight and non insight solutions in LH. The gamma burst could not be related to the motor response, because the button press was done bimanually (i.e., should have increased activation in both hemispheres) and both insight and non-insight problems required button presses. This study suggests that semantic integration (occurring in the RH aSTG) is important for connecting various problem elements together and for connecting the problem to the solution, leading to insight, at least for verbal problems.

In a recent attempt to further unravel the neurobiological underpinnings of insight problem solving, Sandkuhler and Bhattacharya (2008) extended their search to four different aspects of insightful problem solving: mental impasse, restructuring of the problem, deeper understanding of the problem, and the suddenness of the solution, all during performance in a compound remote association task. They found neural correlates for mental impasse in parieto-occipital brain regions in the gamma frequency band (selective attention) and theta frequency band (working memory) which, according to the authors, suggested increased top-down control and increased memory search leading to attentional overload. Moreover, functional fixedness of the mental impasse was associated with increased gamma frequency band activity in right parieto-occipital regions. Parieto-occipital gamma band frequencies were also stronger for correct solutions (deep insight) than for incorrect false positive solutions (there was subjective insight but it was incorrect, thus leading to less deep understanding). The right prefrontal brain regions were implicated in the restructuring of the problem, here alpha band frequencies were increased compared to no restructuring conditions, and this result showed consistency with involvement of this brain region in planning open-ended tasks. Suddenness of the solution was likewise related to power in the gamma frequency band (38–44 Hz) at parieto-occipital regions mainly in the right hemisphere, just before resolution response. Again, as was also reported by Jung-Beeman et al. (2004) the right hemisphere appears to be mainly involved in both mental impasse, the restructuring of the problem, and the suddenness of the solution.

2.4.3 Future Directions

Neurophysiological studies in insight problem solving have almost invariably employed (language related) compound association-like tasks. Sometimes, hints were given to the subjects to improve their performance and thus to elicit 'insight'. Of course, such highly controlled artificial conditions do not score high on ecological validity. Moreover, (artificial) insight as elicited under such conditions may be different from real-life insight and may also be governed by different underlying cognitive- and neurophysiological mechanisms. Therefore, in order to complement current neurophysiological knowledge on insight, and to approximate real-life insightful problem solving, studies could be designed where more complex concepts rather than words or numbers have to be 'discovered', for example the rules or mechanism underlying (simple) physical problems (e.g., gravity, momentum) such as employed in inquiry learning. Further, the role of memory (overload) and the role of attention switching should be explored. Another area of research concerns the precise role of the right hemisphere, which can be explored by selectively presenting visual input to the right (LH) or left (RH) visual field. Future research should also clarify the role of the reported RH gamma just before insight occurs: is it a true manifestation of insight or just an epiphenomenon?

2.5 Implicit vs. Explicit Knowledge/Learning

2.5.1 Education

In implicit learning, knowledge is acquired without explicit intention of learning, without awareness of the learning process and without knowledge of what has been learned. This type of incidental learning differs from explicit learning, which is conscious and intentional. Up to now, little attention has been given to implicit learning in the educational literature. Much work has been done on a related topic, informal learning, which basically is the learning that takes place outside the organised, official, schooling institutions (for example in a museum). Though this type of learning can be conscious and intentional, it may also share elements with implicit learning.

There are two major views with respect to implicit learning. One line of thought claims that rules can be abstracted implicitly, as, for example, has been shown in experiments on artificial grammar learning (AGL) where participants are instructed to memorize stimuli structured by a rule, and are later able to classify stimuli into regular and irregular items with above-chance accuracy without being able to verbalize the rule (see e.g., Reber, 1989). A competing vision states that subjects do not learn abstract rules, but that they are sensitive to frequently occurring features, and that they extract probabilistic information about the composition of sequences or procedures (Shanks & StJohn, 1994). In the case of implicit sequence learning, the serial reaction time task (SRTT) is usually adopted, demonstrating implicit learning when subjects respond faster in response to reoccurring sequences than to random sequences, without being aware of these reoccurrences.

Although the role of implicit learning is recognized in several aspects of knowledge acquisition (e.g., native language learning, second language acquisition), its contribution to education is relatively small in comparison to the impact of explicit learning. This was recently demonstrated in a study by Saetrevik, Reber, and Sannum (2006) who employed an implicit learning paradigm for teaching atomic bonding rules in chemistry. Several conditions were tested, among which a simple one consisting of mere exposure to correct bonding models, gradually elaborating on the rule that governed the bonding of the carbon atom through memorization, counting the atoms, counting the bonds and verifying the bonds. Classification was tested and subjects were asked about explicit knowledge, showing above chance performance even for subjects that were not given explicit information about the rule, thus demonstrating implicit learning, but performance was far better for explicitly instructed subjects.

2.5.2 Cognitive Neuroscience

Quite a few neuroimaging studies, including those conducted with patients (e.g., Amnesia, Huntington), have been conducted on implicit learning. This work shows that separate cortical and subcortical brain regions underlie memory mechanisms in implicit (usually procedural) and explicit (usually declarative) learning. Medial temporal lobe (MTL), the anterior cingulate cortex (ACC) and the medial prefrontal cortex (MPFC) are the brain regions implicated in explicit learning and declarative memory, while striatal (basal ganglia, caudate nucleus) structures have been found to subserve procedural memory and implicit learning (Destrebecqz & Cleeremans, 2001; Reber, Gitelman, Parrish, & Mesulam, 2003). However, it was recently suggested that implicit and explicit learning may share the MTL memory system to some extent (Rose, Haider, Weiller, & Buchel, 2004; Schendan, Searl, Melrose,

& Stern, 2003). In a recent study by Aizenstein and colleagues (Aizenstein et al., 2004), explicit and implicit sequence learning led to learning in both conditions and activation in the prefrontal cortex (PFC), striatal regions, the ACC, and several visual regions. Interestingly, these authors found different activation patterns in the visual cortex in response to the implicit-explicit manipulation, with decreased activation after implicit learning and increased activation in the explicitly learned patterns, but common striatal activity. Reber and colleagues (Reber et al., 2003) also reported differential occipital visual activation for an implicitly learned categorization rule, but found increased activation relative to the implicit learning condition in several brain regions, including the hippocampus, left inferior temporal cortex and posterior cingulate, for explicit intentional learning. In another study, Destrebecqz et al. (2005) found activity in the striatum during recall of an implicitly learned sequence, while ACC/MPFC was recruited for explicit learning. Interestingly, Destrebecqz et al. (2005) report a functional connection between the ACC/MPFC and the striatum (caudate nucleus) during recall after explicit learning, whereas these two systems appear disconnected during recall after implicit learning, thus complementing the reported overlap with respect to implicit and explicit learning in the implicit memory-system dedicated striatum (Aizenstein et al., 2004).

Neurophysiological differentiations between implicit and explicit learning have also been found when children were compared to adults, as was reported by Thomas et al. (2004), who studied developmental differences in the striatum during implicit learning. Thomas et al. (2004) found that neither adults nor children became explicitly aware of an implicitly learned sequence, but that there were differences between the groups with respect to speed and magnitude of the implicit learning effect, where the adults outperformed the children. Interestingly, adults showed more activity in cortical motor regions whereas children displayed more activity in subcortical motor structures (bilateral putamen). Learning-related developmental differences were reported for the hippocampus and superior parietal cortex, but learning related activity in the (right) caudate did not vary with age.

In the aforementioned studies, SRTTs were employed with concurrent neuroimaging (PET; fMRI), but interesting indicators of implicit learning were also found in the language domain using event-related-potentials (ERPs). In a study on second language learning in native speakers of English who learned Spanish as a second language, Tokowicz and MacWhinney (2005) showed that learners were sensitive to violations (i.e., showed different brain responses to grammatical and ungrammatical sentences; P600) in the second language (L2) for constructions that are formed similarly in the first language (L1), but were not sensitive to violations for constructions that differ in the L1 and the L2. Also, a grammaticality effect was found for the construction that was unique to the L2, suggesting that the learners were implicitly sensitive to these violations. Behavioural data showed that judgment accuracy was near chance for all constructions. These findings suggest that learners are able to implicitly process some aspects of L2 syntax even in early stages of learning, but that this knowledge depends on the similarity between the L1 and the L2. In a similar vein, but now concerned with semantics rather than syntax, Thierry and Wu (2007) used an implicit priming paradigm to assess whether Chinese-English bilinguals spontaneously access Chinese translations when reading (or listening to) English words. The authors found that implicit priming had no behavioural effect, but that it modulated the N400 ERP component, suggesting implicit access to meaning in the first language, although the bilinguals read words in their second language.

2.5.3 Future Directions

As was indicated above, under some conditions, implicit and explicit learning and their procedural and declarative memory systems appear to overlap; future research should shed light on this apparent intricate interplay.

Sleep, and particularly slow wave sleep, is important for memory consolidation (Backhaus et al., 2007; Marshall, Helgadottir, Molle, & Born, 2006), but there appear to be differences in the beneficial effects of sleep when implicit learning is compared to explicit learning. Rapid eye movement sleep (REM), rather than slow wave sleep seems to improve implicit learning (Marshall & Born, 2007). Non-REM (NREM) sleep, on the other hand, does not seem to have an advantageous effect on implicit learning (Robertson, Pascual-Leone, & Press, 2004), although others do report positive effects of NREM (stage 2) sleep (Peters, Smith, & Smith, 2007). The issue is even more complicated by the fact that there appear to be developmental differences (Fischer, Wilhelm, & Born, 2007). Clarification, therefore of the precise role of (types of) sleep in memory consolidation of implicitly learned material seems warranted.

Another relevant topic is related to the question of whether or not there is transfer from implicit to explicit knowledge. Lang et al. (2006), for example, showed that several ERP components, including very early ones, differentiated solvers from non-solvers in an implicit sequence paradigm. For some of these solvers, this even led to conscious awareness (insight) and thus explicit knowledge. Future research should further explore the underlying mechanisms and the relation between implicit learning and (precursors of) insight.

Further, the role of prior knowledge and expectations in implicit learning has been acknowledged (see e.g., Sun, Merrill, & Peterson, 2001) and should be further explored, because in usual everyday situations, in contrast to the artificial paradigms employed in the laboratory, such knowledge and expectations may play an important role, not only (for obvious reasons) in explicit learning, but also in implicit learning. Prior knowledge might be a manifestation of earlier experiences with or exposures to the learning material, and might, on the basis of similarity, be associated with the implicit memory of this material (Ziori & Dienes, 2008).

Rather than compare implicit learning with explicit learning, insight into implicit learning mechanisms could also benefit from studying the differences within (more or less successful) implicit learners, see, for example, (Reiss et al., 2005).

A further relevant line of research to be pursued with respect to implicit learning concerns the unconscious detection of errors as reflected in error-related negativity (ERN), a component of the ERP which is seen when errors are made, or feedback of an error is given (see e.g., Nieuwenhuis, Ridderinkhof, Blom, Band, & Kok, 2001), and whether and how this contributes to implicit learning (see also Ferdinand, Mecklinger, & Kray, 2008).

2.6 Metacognitive and Regulative Skills

An important aspect of recent educational developments is the emphasis on self-regulation and self-direction on the part of the learners (Bransford et al., 1999). It is seen as important that learners develop these skills in order to be able to cope with the huge amounts of information available nowadays and in order to be able to continue their learning after formal schooling (life-long learning). This development marks a shift from teacher-controlled to learner-centred instruction (Duffy & Jonassen, 1992; Jonassen, 1999). The result of this shift has fuelled the development of learning paradigms like whole-task learning (van Merriënboer & Kirschner, 2007), where learning is driven by work on rich learning tasks based on whole tasks with a high degree of authenticity. Similarly, inquiry learning, has students explore a domain, usually in science, develop questions in the process of investigating domain aspects, and then test those questions to develop new understanding (de Jong, 2006). Self-regulated learning implies that more of the planning, monitoring, and evaluating of the learning process is in the hand of the learner. Since learners are not always capable of this, adequate learning environments provide learners with support in this respect (Manlove, Lazonder, & de Jong, 2007).

2.6.1 Education

An important higher-order skill is self-regulation (Flavell, 1971), that is, the ability to regulate ones own learning process. It includes three essential activities: planning, monitoring, and evaluation (Butler & Winne, 1995; Schraw, 1998; Zimmerman, 2000). Planning involves goal setting and determining strategies for goal attainment. Monitoring and evaluation involve judgements of how well and to what degree a plan is successfully executed, with monitoring occurring *during* the execution of a plan (e.g., task performance), and evaluation occurring at an end or stopping point (e.g., after task performance, Schön, 1991), providing input for the next plan.

For self-regulation to lead to improvements in learning, the accuracy of 'feelings of knowing' and 'judgements of learning' made during monitoring and evaluating is important (Koriat, 2000; Metcalfe, 1986; Thiede, Anderson, & Therriault, 2003). That is, without accurate assessments of comprehension or performance, students cannot decide whether they have to restudy something or engage in new planning to correct their errors. Unfortunately, learners are often not very accurate at assessing the extent to which they learned something (see e.g., Thiede et al., 2003), or at identifying errors when they are not 'cued' in some sense by their environments to do so (e.g., through feedback, Butler & Winne, 1995). An implicit assumption that seems to be made in educational research is that monitoring and evaluation require conscious reflection to be accurate, and therefore, many attempts have been made to stimulate students' ability to reflect (see e.g., Boud, Keogh, & Walker, 1985; Ertmer & Newby, 1996; van den Boom, Paas, van Merriënboer, & van Gog, 2004). An important question addressed in cognitive psychology, however, is whether certain metacognitive processes are actually implicit or explicit, in other words, whether or not they require awareness (Koriat, 2000; Reder & Schunn, 1996). The answer to this question may have serious implications for the way in which education and instruction can evoke and support metacognitive processes.

2.6.2 Cognitive Neuroscience

In neurosciences, higher-order processes including self-regulation are often referred to as executive control processes. Executive control is an umbrella term for a number of component functions, including selective attention, conflict resolution, error detection, and inhibitory control, which is the cognitive ability to suppress a dominant, though task inappropriate, response in favour of a more goal-appropriate response (Fernandez-Duque, Baird, & Posner, 2000; Shimamura, 2000).

Conflict resolution might play a role in performance monitoring, for example when learners try to resolve incongruence between plans, comprehension, a current state of an activity and (either internal or external) feedback they receive. fMRI studies with the Stroop task wherein a participant is asked to name a word colour, are often used for imaging studies of conflict (e.g., Bernhardt, 1991; Carter, Mintun, & Cohen, 1995). In this task, colour words are printed in their corresponding ink colour (congruence) and in different colours (incongruence). Participants must inhibit the dominant response of naming the word itself in favour of the less dominant response of naming the colour. When the ink colour and the colour word are incongruent, consistent activation patterns have been found indicating common areas involved in conflict resolution. Fernandez-Duque et al. (2000) indicate that '... in the congruent trials metacognitive knowledge (i.e., awareness) of conflict appears to be absent even though there is evidence of metacognitive regulations (i.e., selection of ink colour and filtering of word meaning). This result, if confirmed, would provide convergent evidence for the existence of implicit metacognitive regulation.' (p. 292). Hence, such findings might contribute to answering the question of the degree to which metacognition is implicit or explicit in its functioning, which has important consequences for the effectiveness of educational measures that seek to enhance metacognitive processes. Conflict resolution also occurs in conceptual change when learners have to decide to change or maintain their current ideas on the basis of conflicting information as may come out of an experiment conducted in, for example an inquiry learning session (Chinn & Brewer, 1993). Petitto and Dunbar (in press) investigated conceptual change issues with regard to neurological patterning in an fMRI study that investigated how students make changes to their understanding of concepts they find plausible or implausible. Conceptual change refers to the idea that previously held knowledge which is considered naïve or incorrect on the part of students can be changed through instructional interventions such as presentation of anomalies or deviations from these ideas (e.g., Baker & Piburn, 1997). Conceptual change has been particularly hard to assess or observe. Fugelsang and Dunbar (2005) therefore investigated networks in the brain which were activated when students learn scientific knowledge. Fugelsang and Dunbar (2005) hypothesized that data inconsistent with plausible theory would be ignored and not result in changes to concept understanding, whereas data consistent with plausible theory would be integrated with the given concept. They found that people given data consistent with their theories activated networks involved with learning (caudate and parahippocampal gyrus). However when presented with data that were inconsistent with preferred theory, areas involved in conflict resolution, i.e., anterior cingulated cortex, and dorsolateral prefrontal cortex (DLPFC) are activated. This indicates that shallow presentation of anomalies might not promote conceptual change, since learning areas were not activated, and may show that students actually inhibit or ignore this information as the authors hypothesized. In contrast, when students were presented with extensive data inconsistent with theory, fMRI data does show evidence of learning network activation.

Regarding *error detection*, research has shown that performance slows down following the detection of an error (Robertson, Manly, Andrade, Baddeley, & Yiend, 1997). This strategy adjustment observation lead neuroscientists to propose an error monitoring system located in the medial areas of the frontal lobe, particularly the anterior cingulate, which shows increased activation in response to errors (Carter et al., 1998; Fernandez-Duque et al., 2000). As indicated in Section 2.5.3, event-related potential (ERP) research has identified an ERP component called error-related negativity (ERN, Badgaiyan & Posner, 1998; Ferdinand et al., 2008) which is seen when we make a mistake, or get the feedback that we made a mistake. Interestingly, it also occurs when we observe a mistake being made (see also Section 2.7), and research has shown that this ERN is sensitive to the degree of an error and its subjective meaning. Moreover, it may also occur in response to implicit errors, that is, errors that participants are not explicitly aware of (Ferdinand et al., 2008; Nieuwenhuis et al., 2001). Such neuroscience research might provide us with an understanding of the biological basis of how we detect errors, which plays a crucial role in self-regulation.

It is also important to consider the developmental perspective in this context, because a network of frontal brain areas seems to play a crucial role in executive control (Fernandez-Duque et al., 2000), and the maturation of some of these areas (e.g., prefrontal cortex) continues well into early adulthood. Hence, self-regulated learning, or rather, its effectiveness, might be dependent upon the stage of brain/neurocognitive maturation (see also Section 7.2). That is, children or young adolescents might not be able to engage in self-regulated learning effectively, because the necessary brain areas may not have fully matured. On the other hand, it might also be that younger learners engage different brain areas or use similar brain areas in a different manner (cf., Crone & Huizinga, 2006) especially in the case that explicit awareness is not required (see above).

Metacognition, in the sense of knowing what you know, is related to regulative processes. Schnyer, Nicholls, and Verfaellie (2005) investigated the brain areas involved in feelings of knowing (FOK). Theoretically, the FOK paradigm assumes that FOKs are made based on the relative familiarity of the recall cue (Schnyer et al., 2005). Their results show, however, that the right ventral medial prefrontal cortex (VMPC) was activated during *accurate* retrieval judgments, regardless of actual recall or anticipated recognition of a target item. They go on to conclude that the VMPC's function might be less related to memory retrieval and more to an intuitive assessment of 'feeling of knowing', that is, to monitoring. As we have seen FOK also plays a role in solving insight problems (see Section 2.4).

2.6.3 Future Directions

The research discussed here indicates some areas where neuroscience research may provide important contributions to our theoretical understanding of monitoring and evaluation processes' prerequisites for self-regulation. An important question that neuroscience research might help answer is to what degree metacognitive monitoring is implicit or explicit (cf., Fernandez-Duque et al., 2000). Another important question that neuroscience research might help answer concerns the development of executive control functions, which will provide information with regard to optimal timing of educational endeavours, such as how much regulation we can expect from children in comparison to teenage or adult learners.

2.7 Social Cognition and Social Learning by Observation and Imitation

Educators recognize the importance of social processes for learning. Influential theories in this area are those of Vygotsky (1978) and Bandura (1977; 1986). For example, in Vygotsky's work, social interaction is held to play a fundamental role in (development of) cognition, and Bandura's social learning theory stresses the importance of observing and modelling the behaviours, attitudes, and emotional reactions of others for learning. Especially in vocational training, a large and important part of our educational system, much of the training is performed 'in situ'. Students learn in a (cognitive) apprenticeship mode (Collins, Brown, & Newman, 1989) in which part of the learning takes place by observing experts.

2.7.1 Education

Social learning, that is, learning by observing and imitating others, has long been recognised as a powerful learning strategy for humans (Bandura, 1977, 1986; Collins et al., 1989; Vygotsky, 1978). The terms observational learning and imitation learning are often used interchangeably. However, they can be differentiated as learning can occur without imitation, that is, we may learn by observing and generating inferences beyond the observation without imitation.

In evolutionary psychology, it is argued that we may have *evolved* to observe and imitate other people (see, Sweller & Sweller, 2006). This seems to apply in particular to what Geary (2007) refers to as biologically primary knowledge, that is, knowledge that we have evolved to acquire almost automatically (e.g., face recognition, first language). However, also in acquiring biologically secondary knowledge, which has to be explicitly taught (e.g., writing, arithmetic), learning from expert models has been shown to be very effective (see, Renkl, 2005; Sweller et al., 1998). Learning from expert models can be done by observing the model directly, either 'live', as in a cognitive apprenticeship construction, or on video. But it can also be indirect, through worked-out examples that make the solution steps an expert performs explicit (e.g., in solving a mathematics problem). These instructional strategies rely (in part) on observation/imitation learning, and are used for teaching both motor tasks and cognitive tasks.

2.7.2 Cognitive Neuroscience

An interesting finding from cognitive neuroscience for social learning is the discovery of the mirror-neuron system (for a review, see Rizzolatti & Craighero, 2004), which is thought to play an important role in the understanding of actions made by others, and, hence, in our ability to learn by observing and/or imitating others.

It has been shown that observing (object-oriented) actions made by others activates the mirror-neuron system, which is also active when one performs that action oneself (Iacoboni et al., 1999; Meltzoff & Prinz, 2002). Several authors (Buccino et al., 2004; Craighero, Bello, Fadiga, & Rizzolatti, 2002; Vogt, Taylor, & Hopkins, 2003) found that the mirror-neuron system, which is active during mere action observation, primes the execution of similar actions, and thereby mediates imitation-based learning. For a while, it was thought that the mirror neuron system was only activated when the parts of the human body that executed the action were visible, and not when the action was conducted by some other agent such as a robot arm (Tai, Scherfler, Brooks, Sawamoto, & Castiello, 2004). However, recent evidence suggests that the goal of the observed action is more important for activation than, for example, the presence of a human or robotic hand (Gazzola, Rizzolatti, Wicker, & Keysers, 2007).

These findings may help educational researchers understand the biological bases of observational learning, and provide some insights into why certain instructional designs are more effective than others. For example, van Gog, Paas, Marcus, Ayres, and Sweller (2009) have noted that the mirror neuron system may also contribute to our understanding of an unresolved issue in educational research, specifically, why sometimes dynamic visualizations are more effective than static ones, but sometimes static ones are more effective than dynamic ones (for reviews, see Höffler & Leutner,

2007; Tversky, Morrison, & Betrancourt, 2002). van Gog et al. argue that dynamic visualizations involving human movement may have benefits over static visualizations, because they activate the mirror neuron system. Other types of dynamic visualizations that depict natural, mechanical, or abstract processes do not have this benefit, which may explain why in these cases they are equally or even less effective than static visualizations. Of course, joint research ventures would be necessary to investigate this hypothesis.

However, it should be noted that the above applies to learning (psycho)motor skills, that is, skills that involve human movement. An important open question is whether the mirror neuron system can also explain why instructional formats such as worked examples are effective for teaching cognitive skills. Interestingly, there are indications that the mirror neuron system also becomes active when people listen to sentences that describe the performance of actions by humans, with, for example, hand, mouth, or leg (Tettamanti et al., 2005). An interesting question for – joint – future research would be to investigate whether this would also apply to hearing or reading sentences regarding purely cognitive actions.

It has also been suggested that the mirror neuron system may play a broader role in social cognition by enabling understanding of actions made by others, i.e., there might be a link with empathy and development of theory of mind (see e.g., Keysers & Gazzola, 2007). However, it is questionable whether the mirror neuron system alone is involved here (this, by the way, can also be asked regarding the findings on imitation described above). That is, there may be a complex interplay between neural circuits involved in motor control, mental simulation, and mirroring that enable imitation and empathy (see e.g., article by Hurley, 2008, and commentaries). Recent findings indicating that the development of self-evaluation and social monitoring may not take place before middle adolescence in the majority of youth (e.g., Amodio & Frith, 2007; Paus, 2005; Steinberg, 2005) are of major importance in this regard. It has been hypothesized that the ability to cognitively evaluate (i.e., mentally simulate) action programs in terms of emotional consequences and social consequences is dependent upon the development of self-evaluation and social monitoring. In other words, the adolescent brain learns to prioritize competing action programs (and parts thereof) in terms of the consequences which these actions have in the short term (e.g., in the next minutes or hours), or medium term, semi long term (e.g., weeks or months), or long term (e.g., years) and the consequences these actions have for 'significant others' (peers, friends, parents, teacher) and society, including social norms (see also Section 7.2 where more general aspects of maturation are discussed).

2.7.3 Future Directions

It should be noted that although the findings regarding the mirror neuron system are promising, the types of tasks used are often very simple, for example, playing a guitar chord (Buccino et al., 2004) or grasping an object (Gazzola et al., 2007). The question remains whether these findings also hold for more complex motor tasks. In addition, as mentioned above, it is unclear what these results can tell us about observational and imitation learning of cognitive and linguistic tasks, although the findings of Tettamanti et al. (2005) seem promising in this regard. Joint research ventures are necessary on educationally relevant motor tasks, cognitive tasks with or without motor components, and instructional design implications. For example, regarding the design of instructional visualizations, future research should make careful comparisons between dynamic and static visualizations of human movement and other types of instructional animations on activations of the mirror neuron system (van Gog et al., 2009). In addition, educational implications of (the development of) social cognition in general should be addressed.

Chapter 3
Affective Processes in Learning

3.1 Education

In the last decade the role of emotions in education seems to have been rediscovered (Maehr, 2001). Affective processes are now recognised as playing an important role in learning. Students' emotions, such as, enjoyment, boredom, pride, and anxiety are seen to affect achievement by influencing the student's involvement and attitude towards learning and learning environments, which also affects how (intensively) students process and/or interpret information (for a discussion see e.g., Boekaerts, 2003; Boekaerts & Simons, 1995; Pekrun, 2005; Pekrun, Goetz, Titz, & Perry, 2002). This (renewed) attention for affect is the result of several developments. A first development is the change from teacher-directed to learner-centred approaches in education, which often involve giving more responsibility for and control over the learning process to the learners. Consequently, affective-motivational processes and self-regulatory processes become more and more important (Boekaerts, 2003). A second, related development is the perspective that the effects of educational methods are often not determined by these methods per se, but by the way they are *perceived* by the learners. Thus, student perceptions, expectations, and appreciations are seen as increasingly important in the study of instructional methods (e.g., Könings, Brand-Gruwel, van Merriënboer, & Broers, 2008). A third development is related to a social-constructivist view on education, with a greater focus on group work and collaborative learning. From this perspective, affects and emotions in group work becomes an important research topic (e.g., Peterson & Schreiber, 2006).

3.2 Cognitive Neuroscience

In basic neuroscience and biological psychology it has been a known fact for decades that the brain areas involved in emotional processing are of prime importance for learning. As early as the 1950s, insights were obtained on the crucial role of limbic structures for memory consolidation and these same structures were also known to be involved in elementary emotional processing. Emotional processing appeared to be necessary for proper memory consolidation to occur, and both animal

T. de Jong et al., *Explorations in Learning and the Brain*,
DOI 10.1007/978-0-387-89512-3_3, © Springer Science+Business Media, LLC 2009

and human research shows the major importance of emotional and motivational processing and involvement of particular neurotransmitters and neurohormones. An overview can be found in Kolb and Whishaw (2008). These research findings have been revitalized by recent brain imaging experiments which suggest the role of limbic areas in temporal lobe and prefrontal (notably anterior cingulate) functioning (e.g., Amodio & Frith, 2007). The study of affective processes is a rapidly emerging sub-field in neuroscience (e.g., the journal *Cognitive, Affective, & Behavioral Neuroscience* was started in 2000, and *Social, Cognitive, and Affective Neuroscience* in 2006).

Several authors have investigated how materials with different emotional valence are processed. For example, Simpson et al. (2000) investigated effects of emotional valence (negative or neutral) of pictures presented during a cognitive task on task performance. They found that task performance was slower for negatively valenced pictures, and that there were differences in the functional anatomy associated with task performance for negative and neutral pictures. Dolcos and Cabeza (2002) showed that emotional events were remembered better than non-emotional events. They measured ERPs while participants rated the emotional content of pleasant, unpleasant, and neutral pictures. They found differences in the ERPs for emotional and non-emotional (neutral) stimuli and subsequent recall was better for pleasant and unpleasant pictures than for neutral pictures. Results by Fox (2002) suggest that not only the emotional valence of the stimuli is important for how these are processed, but also the emotionality of the individual engaged in the task. She found that participants with high levels of trait anxiety showed an attention bias towards fearful faces.

Locke and Braver (2008) studied what changes in brain activity are associated with changes in motivational state. Participants performed a task under three motivational conditions (reward-incentive, penalty-incentive, and baseline). The fMRI data indicated that reward was associated with activation increase in a right-lateralized network including parietal and prefrontal cortex. Interestingly, they also took personality measures and showed that individual differences in activation were related to motivation-related personality variables. Their results suggest that changes in motivational state may modulate performance through sustained activity in cognitive control regions and that the effect of incentives may be affected by the personalities of the participants.

Another study from neuroscience that emphasizes the importance of emotion is the study by Immordino-Yang and Damasio (2007). Based on evidence from brain-damaged patients, they suggest that emotion-related processes are also required for transfer of skills and knowledge from school to real-world environments. That is, patients whose cognitive skills were intact, but whose emotional responses were damaged (i.e., they had problems with social emotions, such as compassion, embarrassment, and guilt), were no longer able to select the most appropriate response in social situations based on their past experience, and were no longer able to learn from feedback, i.e. 'emotional repercussions', on their behaviour. This is taken by the authors as an indication that even though their logic and knowledge was intact 'their reasoning was flawed because the emotions and social considerations that underlie good reasoning were compromised' (p. 5). The authors suggest based on findings with these and other patients that emotional processes may underlie rational decision making and learning.

In addition, there is a strong link between social cognition (see Section 2.7) and emotional processes. Social stimuli function as emotional barometers for the immediate environment and are the catalysts for many emotional reactions (which have inherent value for relationships and survival). Norris, Chen, Zhu, Small, and Cacioppo (2004) conducted a study to test the hypotheses that the neural mechanisms underlying social and emotional information processing could be interconnected. Their study showed that social and emotional processes have both independent and interactive effects on brain activation. Regarding development, as mentioned in the section on social cognition (Section 2.7), it has been suggested that the ability to evaluate action programs in terms of emotional and social consequences which are at stake develops in adolescence.

3.3 Future Directions

The field of affective neuroscience appears to be booming, and it is universally accepted that social and emotional factors exert a strong influence on learning. Therefore, future joint research in this area could provide important contributions to education and educational research. Similar to statements made in earlier paragraphs, effort should be taken to incorporate insights from educational research and practice into (affective) neuroscience in order to come to new paradigms which can help advance educational science through provision of new experimental models with strong explanatory power.

Chapter 4
(Second) Language Learning and Literacy

Literacy is incredibly complex, and a full report on the links between neuroscience and language instruction would be an undertaking all on its own. In this chapter we therefore focus on a number of focal questions: What is the neurological basis of development of literacy? Can cognitive neuroscience help to distinguish between competing models discussed in educational research? What is the role of age of acquisition in second language (L2) learning? Does early L2 learning have a negative impact on acquisition of literacy in the native language? Can late L2 learners process an L2 in a native-like way? First, we give a brief overview of educational research developments in these areas, followed by an overview of neurocognitive contributions.

4.1 Education

4.1.1 Development of First Language Literacy

According to the *2000 National Reading Panel Report: Teaching Children to Read*, reading development is comprised of five essential component skills which build on each other successively; the alphabetic principle, phonemic awareness, oral reading fluency, vocabulary and comprehension (Paris, 2005). The alphabetic principle pertains to the ability to associate phonemes (sounds) to letters and to use these phonemes to read words. Phonemic awareness is the learning of specific phonic units of language including vowels, consonants, and consonant digraphs and their corresponding sounds (Ehri, Nunes, Stahl, & Willows, 2001). Oral reading fluency is considered to follow these prerequisite component skills. In the oral reading fluency stage, the practice of reading becomes automatised which frees up working memory for the final two component skills of vocabulary development and comprehension.

Although general consensus exists with regard to this developmental trajectory, as evidenced by national policies formed by reports such as the *2000 National Reading Panel*, there is some controversy with regard to the developmental order and the importance of decoding skills (alphabetic principle and phoneme awareness)

T. de Jong et al., *Explorations in Learning and the Brain*,
DOI 10.1007/978-0-387-89512-3_4, © Springer Science+Business Media, LLC 2009

over comprehension (Calfee & Norman, 1998). This controversy is mainly found in instructional design approaches to reading such as exemplified in the phonics over whole language debates.

In the literature on visual word recognition (i.e., reading a single word), two types of models are prominent: dual-route models (e.g., Coltheart, Curtis, Atkins, & Haller, 1993; Coltheart, Rastle, Perry, Langdon, & Zeigler, 2001) and models emphasizing the importance of phonological processing (e.g., Frost, 1998; Stone, Vanhoy, & van Orden, 1997). In short, dual-route models propose two distinct routes for reading. One route can be described as direct access from the written word to the mental lexicon which contains the word's meaning and pronunciation. The second route is indirect, in the sense that it requires converting letters to sounds (i.e., graphemes to phonemes) in order to access the word in the mental lexicon. Dual route models propose that beginning readers use the indirect route, in which they (slowly) sound out words, but as reading ability progresses, the direct route will be used more and more frequently. Such models propose that experienced readers would only use the indirect route for reading infrequent words or non-words because the direct route is seen as quicker and more efficient. An increasing number of researchers dispute dual route models in favour of a strong phonological theory of reading, in which phonological processing is mandatory, early, and rapid (e.g., Frost, 1998; Stone et al., 1997).

The controversies with regard to the developmental order in which component skills should be taught, and the importance of comprehension are referred to as 'the reading wars' within the literature on instructional designs that promote literacy. They centre on debates about the phonemic awareness perspective over a whole language comprehension oriented perspective.

Two meta-analysis studies compared these two approaches: Ehri et al. (2001) and Jeynes and Littell (2000). Phonemic awareness instruction, also known as the phonics approach, emphasizes teaching the alphabetic code needed for literacy of written language. Ehri et al. (2001) define this approach as one which incorporates a planned set of phonic elements including correspondence between consonant letters and sounds, vowel and consonant diagraphs (oi, ea, sh, th), and blends of larger sub units in words, such as 'op' in the word stop (p. 394). In contrast, whole language emphasizes '(1) whole pieces of literature and functional language as opposed to abridgments, adaptations, or segmented texts; (2) individual students' choice as opposed to teacher-sponsored, whole-class assignments; and (3) integrated language experiences as opposed to direct instruction in isolated skill sequences' (Jeynes & Littell, 2000, p. 21). These two meta-analysis studies (Ehri et al., 2001; Jeynes & Littell, 2000), one focusing on phonics research, and the other looking at the effectiveness of whole language approaches are suggestive of the dichotomy which is apparent in the literature.

Calfee and Norman (1998) sum up attempts to resolve the question of which approach is more effective in the classroom and conclude that 'The outcome of these investigations seemed clear-cut: (1) Teacher-led direct phonic programs produce (slightly) higher scores on decoding measures at the end of the first grade but (2) variability between teachers within programs was substantial, (3) many students did poorly under all programs, and (4) the initial advantages washed out by the end of third grade' (Calfee & Norman, 1998, p. 243).

Results from both Ehri et al.'s (2001) and Jeynes and Littell's (2000) meta-analyses suggest that phonics instruction has particularly strong effects on low and middle socioeconomic status (SES) readers. Ehri et al.'s 2001 meta-analysis also found that phonics instruction was particularly beneficial to early readers and readers with reading disabilities, in addition to being more effective than whole-language or other (control group) forms of instruction. They conclude, however, with the speculation that the effectiveness of whole language instruction could be enhanced if it were 'enriched' with systematic phonics.

Children's SES is only one of the many factors that may influence the development of literacy and instructional approaches to literacy. Another important factor is the orthographic complexity of the language, which is thought to impact which reading strategies are employed and how quickly or easily literacy develops. The extent to which a language is orthographically complex relates to how a language is structured at a 'grain-size' (e.g., Ziegler, Perry, Jacobs, & Braun, 2001) from decoding single sounds (t), to mixed sounds (th), to whole syllables and whole words (Kanji). Zeigler and Goswami (2006) argue that inconsistency in the symbol-to-sound mapping impacts literacy development. If a language is inconsistent in its pronunciations or has multiple pronunciations it may be more challenging to learn. Zeigler and Goswami (2006) cite cross-language reading comparison research conducted by the *European Concerted Action on Learning Disorders as a Barrier to Human Development*. Fourteen European Union countries participated in assessments of children's reading with word and non-word tests. A striking finding was that children who were acquiring reading in orthographically consistent languages (Greek, Finnish, German, Italian, Spanish) were close to ceiling in both word and non-word reading by the middle of the first grade. Danish (71% correct), Portuguese (73% correct) and French (79% correct) children showed somewhat reduced levels of recoding accuracy, which is in line with a lower orthographic consistency of these languages. In contrast, English speaking children performed extremely poorly (34% correct).

This characteristic of language's impact on literacy development is related, according to Zeigler and Goswami (2006) to a cross-language theory of reading – the orthographic depth hypothesis (ODH) (Frost, Katz, & Bentin, 1987; Katz & Frost, 1992). This hypothesis states that 'different psycholinguistic units develop in response to differences in orthography'. Furthermore, the ODH suggests that readers adapt their reliance on the 'orthographic' (whole word recognition) or 'phonological' (recoding) route, depending on the orthographic depth of their language. In a consistent orthography, readers rely more on the 'phonological' or nonlexical route, because mapping between two letters and sounds is relatively direct and unambiguous. In an inconsistent orthography, readers rely less on the phonological route and to a greater extent on the lexical or 'orthographic' route (Zeigler & Goswami, 2006. p. 434). It should be noted that ODH is based on a dual-route model of word recognition (Coltheart et al., 2001) and, as discussed above, an increasing number of researchers dispute dual route models in favour of a strong phonological theory of reading (e.g., Manis, Seidenberg, Doi, McBride-Chang, & Petersen, 1996; Pennington, 1999).

4.1.2 Second Language Learning

When thinking about language learning and literacy it is important to keep in mind that the world's language system is rapidly changing because of demographic trends, new technologies, and international communication (Graddol, 2004). One of the consequences is that the majority of present and future generations of children will learn more than one language. This is certainly true in the Netherlands, where an increasing number of children are exposed to two (or more) languages at an early age. Furthermore, the Dutch curriculum emphasizes the learning of foreign languages, and English, for example, is taught from 5th grade primary school onwards. The international relevance of this issue was underscored in the 125th anniversary issue of *Science*, in which the editors compiled a list of the 125 critical questions for the next 25 years. One of these questions pertained to the biological basis of second language (L2) learning as revealed by the monitoring of brain activity (Kennedy & Norman, 2005). This research question ensued from *Science*'s observation that 'children pick up languages with ease while adults often struggle to learn train station basics in a foreign language (p. 93)'.

In regard to second language learning, there seems to be considerable overlap in reading skills in the first and second language (Bernhardt, 2000). Bernhardt's (1991) extensive review of the literature indicates that learning to read in the first and in the second language require similar skills. For example, fluency is related to speed of processing, and phonological processing is key to word recognition in all languages, 'even in languages that are non-alphabetic and considered more conceptual in nature' (Bernhardt, 2000, p. 797). An important issue for second-language instruction is the issue of timing. Although children are generally considered to acquire fluency in languages easily, paradoxically, some educational systems hold that exposure to bilingual education too early will impede progress in a first language. Petitto and Dunbar (in press) refer to this as the 'hold-back' position. These authors cite two classes of hypotheses with regard to this bilingual paradox; the unitary and differentiated. In the unitary hypothesis, children exposed to two languages are thought to have a fused linguistic representation which becomes differentiated only after the age of 3 (Redlinger & Park, 1980). In the differentiated position, researchers claim that bilingual children do differentiate between their two languages (Genesee, Nicoladis, & Paradis, 1995).

4.2 Cognitive Neuroscience

4.2.1 Development of Literacy

Cognitive neuroscience research has helped to illuminate some differences between children and adults on reading tasks by investigating brain activation during such tasks (Booth et al., 2000; Gaillard et al., 2000). For example, Booth et al. (2000) showed that both children and adults utilize the left frontal cortex when engaged in silent reading, but that children's fMRI scans showed increased activation patterns during the task in comparison to adults. The study of Gaillard et al. (2000) examined comprehension and found children to have similar activation patterns to adults with regard to temporal and frontal regions of the brain, but increased activation in the inferior occipital and anterior superior temporal areas. Church, Coalson, Lugar, Petersen and Schlaggar (2008) furthered these lines with an fMRI study with children (ages 7–10) and adults (ages 18–32) during high frequency word reading and repetition tasks. Most brain regions showed similar activity across age groups, indicating that children and adults use overlapping mechanisms when processing high-frequency words. However, age group differences were found in a number of posterior regions implicated in adult reading: the left supramarginal gyrus, the left angular gyrus, and bilateral anterior extrastriate cortex. In comparison to children, adults showed decreased activity in the angular and supramarginal gyrus regions, which are hypothesized to play a role in phonology. These results are consistent with an age-related decreasing reliance on phonological processing. Studies such as these will continue to play an important role in understanding how literacy develops and to serve as a reminder that 'brain activation for adults does not necessarily generalize to children' (Berninger & Richards, 2002, p. 145).

Jobard, Crivello and Tzourio-Mazoyer (2003) carried out a meta-analysis with an aim to provide an objective picture of recent neuroimaging studies concerned with cerebral structures underlying word reading. This study, carried out within the framework of the dual route model of reading, revealed that no areas were recruited more by word than pseudoword reading, implying that the first steps of word access may be common to word and word-like stimuli. The anatomical location of this first step in word access appears to be within a left occipitotemporal region (previously referred to as the Visual Word Form Area-VWFA) situated in the ventral route, at the junction between inferior temporal and fusiform gyri. The results of this meta-analysis also indicated the existence of brain regions predominantly involved in each of the two routes proposed by the dual route model. The authors concluded that the indirect route, characterized by graphophonological conversion, seems to rely on left lateralized brain structures such as superior temporal areas, supramarginal gyrus, and the opercular part of the inferior frontal gyrus. The latter two regions appear to reflect a greater working memory load imposed by the indirect route. The direct route is thought to arise from the coactivation of the VWFA and semantic areas, consisting of a basal inferior temporal area, the posterior part of the middle temporal gyrus, and the triangular part of inferior frontal gyrus. The authors concluded that these results confirm the suitability of the dual route framework to account for activations observed in nonpathological subjects while they read. However, their finding that the first step involved in word access which is common to both word and non-word stimuli could also be related to phonological processing and could therefore also be seen as support for the hypothesis that phonological processing is mandatory, early and fast.

Attempts have been made to take a neuroscience perspective on factors which impact literacy. Noble, Tottenham, and Casey (2005) examined neuroscientific evidence for language and reading and attempt to relate it to racial and SES disparities in neurocognitive performance. The authors cite the work of Mezzacappa (2004) which showed that children from higher SES backgrounds generally outperformed lower SES status students with regard to cognitive control (the ability to ignore distraction, allocate attention and hold items in working memory). Noble, Norman and Farah (2005) examined neurocognitive functioning of African American kindergartners from different SES backgrounds using cognitive neuroscience tasks. The authors found that while SES correlated with performance on the test battery as whole, the effects on language and cognitive control systems in particular were quite large. As cited in the above educational research has provided evidence of the impact of SES on the success of literacy instructional approaches such as phonics or whole-language.

Researchers in cognitive neuroscience research have also addressed the question of the effect orthographic complexity has on language learning and development. Paulesu et al. (Paulesu et al., 2000) for example, compared adult readers of Italian and English. Italian has a consistent orthography, so that readers can consistently convert graphemes into phonemes, whereas English has an inconsistent orthography. Paulesu et al. observed that Italian readers were faster in reading words and non-words than English readers, and different regions in the brain were activated during reading. Italian readers showed greater activation in left superior temporal regions associated with phoneme processing. English readers showed greater activations in the posterior inferior temporal gyrus and anterior inferior frontal gyrus, which are associated with word retrieval during reading.

4.2.2 Second Language Learning

One of the research lines in the cognitive neuroscience of second language learning is to examine the impact of first language learning on second-language learning. Nakada, Fujii, and Kwee (2001) conducted fMRI research with ten Japanese volunteers, five of which were literate in English, and ten American native English speakers, five of which were literate to the same degree as their Japanese speaking counterparts in Japanese. The results showed that cognitive processes for reading in a second language are neurologically similar to those employed by the first language. They regard this as evidence for the hypothesis that the second language represents a cognitive extension of the first language. Further research with Chinese has shown that similar areas are recruited when Chinese speakers read Kanji and when they read English, leading Tan et al. (2003) to suggest that the neural systems of second language reading are shaped by the native language.

Another line of studies focus specifically on the time course of achieving fluency in the second language and factors that may influence this (for a review, see van Hell & Tokowicz, in press). An important issue in the acquisition of literacy in a second language is the amount and timing of second language exposure. McLaughlin, Osterhout, and Kim (2004) investigated ERPs during word identification in adult (English-native) learners of French, and observed that these learners discriminated between words and ‘pseudowords’ (i.e., letter strings following orthographic rules in the language) in their second language after only 14 hours of instruction. Interestingly, when measured with traditional behavioural measures, the learners performed at chance level when making overt word-pseudoword judgments. In an ERP-study on a grammaticality judgment task, Tokowicz and MacWhinney (2005) found that adult (English-native) novice learners of Spanish were sensitive to violations of grammatical constructions in their second language that are formed similarly in their native language. In contrast, the second-language learners were not sensitive to violations for grammatical constructions that were formed differently in the second and native language.

Studies on the timing of exposure to L2 provide important insights into the age at which a second language optimally develops (Kovelman & Petitto, 2002; Petitto, Kovelman, & Harasymowycz, 2003). Kovelman and Petitto (2002) found that prior to age 5, exposure to two languages is optimal for the development of both languages. They also found that children exposed to new languages after this critical time can achieve a fundamental grammatical basis in the second language within the first year, but only if second language exposure occurs in multiple contexts beyond formal schooling. In subsequent neurocognitive research, Petitto et al. (2004) investigated visual perception, speech recognition as well as native and non-native phonetic perception in infants with Near Infrared Spectroscopy (NIRS). The authors found activations in classical language areas of both bilingual and mono-lingual babies (Petitto & Dunbar, in press). fMRI research with adults exposed to two languages before the age of five show evidence of differences in activations for the two languages in comparison to adults who are exposed later in life. This research provides some evidence for psycholinguistic findings that state that language processing declines if the language is learned after puberty (Kim, Relkin, Lee, & Hirsch, 1997; Petitto et al., 2004; Petitto & Dunbar, in press; Wartenburger et al., 2003). For an extensive review, see Abutalebi, Stefano and Perani (2005).

4.3 Future Directions

Currently cognitive neuroscience aims to refine and explain how the human brain decodes words and sentences in native and second languages. Cognitive neuroscience research potentially provides important insights needed to fine tune theories of the developmental trajectories in language learning and the acquisition of literacy. Cognitive neuroscience may also give support for the hypothesis that a balanced approach between phonics and reading for meaning (whole-language) is a key instructional strategy. Finally, neurocognitive research provides crucial insights into the brain processes involved in the learning of foreign languages, which is of particular importance given the emphasis on the attainment of literacy in foreign languages in the Dutch curriculum, and the neurological implications of exposure to multiple languages at an early age.

Chapter 5
Numeracy and Mathematics Learning

Because numeracy, like literacy, results from the interplay of biology and experience it is the natural domain of both cognitive neuroscience and educational science. Although there is no single agreed upon definition, numeracy implies an understanding of the concept of number and the ability to reason quantitatively. As such, it is considered the basis of both simple and complex mathematics.

5.1 Education

In order to design curricula that help children maximize their innate cognitive capacities, it is necessary to first have a detailed understanding of what those cognitive capacities are. Recent research has shown that numerical abilities, like linguistic abilities, are innate to humans, based on the finding that even infants possess certain numerical abilities; for an overview on the development of arithmetical abilities, see Butterworth (2005). Although studies in the 1980s and 1990s concluded that infants are able to make numerosity discriminations between, for example, two and three dots (Starkey & Cooper, 1980) and to perform simple arithmetic operations, such as 1 + 1 (Wynn, 1992) there has been some criticism citing that these studies did not properly control for continuous variables that covaried with numerosity, such as total filled area. Results of more carefully controlled studies show that although infants possess numerical abilities, these abilities appear to be restricted to large numerosity discrimination between, for example, 8 and 16 sounds or dots (Lipton, 2005; Xu & Spelke, 2000). These studies also demonstrate the imprecision of infants' numerosity discrimination, showing that infants are unable to distinguish 8 from 12 sounds or dots (Lipton, 2005; Xu & Spelke, 2000).

It has been proposed that there are two different cognitive systems to assess numerosity: One system for the exact representation of small numbers of objects and another system for representing approximate numerosity (Carey, 2001; Feigenson, Dehaene, & Spelke, 2004). The second system, sometimes referred to as the analogue magnitude system, is thought to be activated during symbolic numerical and mathematical operations using Arabic digits or number words (Dehaene, 1996; Dehaene, Dupoux, & Mehler, 1990).

T. de Jong et al., *Explorations in Learning and the Brain*,
DOI 10.1007/978-0-387-89512-3_5,

In recent decades there has been a shift in many countries from direct instruction, which relies largely on drill and practice, to more realistic mathematics education based on constructionist principles. In the Netherlands, this has taken the form of Realistic Mathematics Education (e.g., Streefland, 1986; Treffers, 1993). However, there is evidence that children with intellectual impairments achieve better results with direct instruction (e.g., Kroesbergen & van Luit, 2005). In addition, because realistic mathematics education places more demand on the skills that students with mathematical learning problems may perform more poorly in, such as vocabulary, reading level, and math fact fluency, there is some question as to whether this method is appropriate for these children (Ruijssenaars, van Luit, & van Lieshout, 2004). It should be noted, however, that there are a number of researchers in the Netherlands that advocate the use of realistic mathematics education with children in special education (Boswinkel, Baltussen, Hoogendijk, & Moerlands, 2003). Clearly, more research is needed to resolve this important issue.

5.2 Cognitive Neuroscience

Some progress has been made in the investigation of the neural substrate of mathematical processes. Evidence from both lesion and brain-imaging studies suggest that areas in the parietal cortex of the brain are involved in number processing (Dehaene et al., 2003), and appears differentially activated by the numerical concepts of 'how many' (counting) and 'how much' (assessing) (Castelli, Glaser, & Butterworth, 2006). Specifically, the horizontal segment of the intraparietal sulcus (HIPS) in both hemispheres is systematically activated during tasks that require access to a semantic representation of magnitude, such as estimation or subtraction. Dehaene and colleagues (2003) suggested that the (bilateral) HIPS might constitute a genetically-defined brain structure for numerical cognition, because of its crucial role in the formation and manipulation of mental magnitude representations and because damage to this area has devastating effects on mathematical abilities. The left angular gyrus (in the parietal lobe), which is part of the language system, is activated during operations such as multiplication that call upon a verbal coding of numbers. In addition, the (bilateral) posterior superior parietal area is associated with visuospatial processing and is thought to be involved in attentional orientation on the mental number line, which implies that this area would be activated during calculations such as subtraction. In addition to these brain areas that appear to be directly involved in numerical cognition, a number of studies have cited the importance of such cognitive processes as executive functions (Mazzocco & Kover, 2007) and working memory (Adams & Hitch, 1997) in mathematics. Whereas executive functions are associated with both prefrontal and posterior (mainly parietal) regions (Collette, Hogge, Salmon, & van der Linden, 2006), working memory has been associated with a dynamic fronto-parietal network ((D'Esposito, Postle, Jonides, & Smith, 1999). An understanding of the neural substrate involved in numerical cognition can contribute to an understanding of the effects of training and instruction.

Ischebeck and colleagues (Ischebeck et al., 2006) showed that training with either multiplication or subtraction led to decreases in activation in inferior frontal areas, indicating that training reduces demand on working memory and executive control. Training in multiplication also led to a shift in activation to the left angular gyrus, suggesting that training caused a shift in strategy from calculation to more automatic retrieval.

5.3 Future Directions

A more thorough understanding of the development of mathematical abilities from a cognitive neuroscience perspective has the potential to facilitate the design of research paradigms in educational research. For example, some children with mathematical learning difficulties seem to make use of immature and inefficient strategies (Geary, 1994). Such strategies can be considered inefficient in the sense that they place greater demands on cognitive processes such as working memory. However, strategy use is commonly determined by verbal or written reports from the children themselves and there is reason to believe that such reports may not accurately reflect strategy use (Kirk & Ashcraft, 2001). The possibility exists that brain imaging could be used as an objective measure that combined with more qualitative data could give indications of strategy use, facilitating research into the effectiveness of strategy and the effects of strategy training. Such a line of research may help answer the question of the effectiveness of realistic versus direct mathematics instruction, particularly for children with limited intelligence and/or specific cognitive impairments. The degree to which specific tasks place demands upon brain areas that mediate not only mental calculation, but also such functions as working memory and executive function could be assessed with brain imaging techniques in both unimpaired children and in clinical groups.

Chapter 6
Learning Problems

6.1 Dyslexia

Many children experience difficulty learning to read although they receive sufficient reading instruction. Reading difficulties that do not result from global intellectual deficits or a chronic problem of motivation are termed dyslexia. Quantifying the (world wide) prevalence of dyslexia would require a universally accepted definition and screening process, both of which are lacking. However, estimates range from 2% to 10% of the population, with higher rates found in languages with a deep orthography. Both within the fields of neuropsychology and cognitive (neuro)science there is considerable knowledge on dyslexia. The present chapter focuses on some issues which are particularly relevant for education. The reader is referred to recent reviews and reports for more in-depth evaluation such as the OECD report (2007) and a paper by Goswami in Nature Reviews Neuroscience (2006).

6.1.1 Education

A number of types of dyslexia have been distinguished; including surface dyslexia, phonological dyslexia, and deep dyslexia (Pennington, 1999). Children with surface dyslexia read as beginning readers do; they break even frequent words into their syllabic constituents and have particular difficulty reading irregular words correctly. These children seem to make use of the assembly of phonology without being able to address the word's phonological structure from the mental lexicon (Coltheart, Masterson, Byng, Prior, & Riddoch, 1983; Shallice, Warrington, & McCarthy, 1983). Children with phonological dyslexia, however, can read familiar words but their reading of novel words or pseudowords is severely impaired (Funnell, 1983). Deep dyslexic readers make semantic errors consisting of mispronunciations that, rather than being phonologically related to the printed word, are semantically related (e.g., *flower* and *rose)*. This seems to suggest that deep dyslexic readers bypass the word's phonological structure and, albeit sometimes inefficiently, access meaning directly from print (Marshall & Newcombe, 1981).

T. de Jong et al., *Explorations in Learning and the Brain*,
DOI 10.1007/978-0-387-89512-3_6, © Springer Science+Business Media, LLC 2009

Some authors (Coltheart et al., 1993; Coltheart & Rastle, 1994) have interpreted these different types of reading difficulty as support for a dual route model of printed word recognition. Children with surface dyslexia would be seen as having a deficit in the direct route, which implies that they must use the slower indirect route of mapping letters to sounds to read even frequent words. Children with phonological dyslexia, on the other hand, are seen as having a deficit in the indirect route and therefore are able to read familiar words but are unable to sound out novel words or pseudowords. Finally, deep dyslexics are also assumed to have a deficit in the indirect route, impairing their ability to sound out words, along with some impairment of the direct route, resulting in sometimes inaccurate direct access to meaning.

Other researchers emphasized the role of phonology in dyslexia (e.g., Manis et al., 1996; Pennington, 1999). Manis and colleagues (1996) explain phonological dyslexics as having degraded phonological representations, which has a maximal impact on pseudoword reading, a somewhat lesser impact on irregular word reading and little impact on regular word reading. Depending on the degree of phonological representation degradation, such children would be classified as phonological or mixed dyslexics. These authors found that the performance of surface dyslexics was similar to that of younger typically developing children, suggesting a developmental delay in word recognition. They explain this delay as being due to a reduced number of units in the middle layer of the connectionist network. As a result, such children can learn rule-like regularities (although less efficiently), but are impaired in learning word-specific patterns.

Another account also acknowledges the role of phonology, but sees it as stemming from a more basic lower level deficit in temporal processing. According to Tallal (1980), reading disabled children have difficulty in processing brief auditory cues or rapidly changing acoustic events in speech and non-speech. As a result, they have difficulty in judging the temporal order (temporal deficit hypothesis) not only of brief rapidly presented non-speech tones, but also of stop consonant-vowel syllables that contrast in their original formant transitions.

Recently, Goswami et al. (2002) have suggested a basic auditory processing theory pertaining to perceptual centres ('P-centres'). Changes in the rate of amplitude modulations in acoustic signals characterize such P-centres, and peak increments in mid-band spectral energy correspond to vowel onset in speech sounds. P-centres can be used to disentangle sub-syllabic segments of onset (the phonemes preceding the vowel) and rime (the vowel and any following phonemes), and processing of aspects of these variables (e.g., onset rise time), which are important for the development of phonological representations, appear to be impaired in dyslexics, see e.g., Richardson, Thomson, Scott, and Goswami (2004).

In addition to deficits in auditory processing, deficits have also been postulated in the visual domain. The so-called magnocellular deficit hypothesis assumes a selective impairment of the visual transient system in dyslexics, leading to impaired sensitivity for low contrast (Lovegrove, Bowling, Badcock, & Blackwood, 1980) or coherent visual motion (Talcott, Hansen, Assoku, & Stein, 2000). The motion detection system is important for the direction of visual attention, eye movements, and visual search, and thus is presumably also involved in the reading process

(Stein, 2001). The magnocellular deficit hypothesis may even be (pan-sensorially) expanded from the visual modality to the auditory modality, and might cause the deficits in auditory temporal processing (Stein, 2001; Tallal, 1980). A recent alternative explanation for the visual abnormalities evidenced by dyslexics is claimed by Sperling, Zhong-Lin, Manis, and Seidenberg (2005) who argue that dyslexics may have elevated (visual) contrast thresholds only when stimuli (not just magnocellular, but also parvocellular) are presented with high noise; they perform similar to non-dyslexics under noise-free conditions.

6.1.2 Cognitive Neuroscience

Several structural differences have been found between the brains of dyslexics and normal controls. For example, cortical microscopic anomalies (Galaburda et al., 1985), the absence of asymmetry in the planum temporale (Galaburda et al., 1985), differences in the corpus callosum (e.g., Rumsey et al., 1996), in the cerebellum (e.g., Rae et al., 2002), and differences in the magnocellular layers of the thalamus, both in the auditory part (medial geniculate nucleus; MGN, Galaburda et al., 1994) and the visual part (lateral geniculate nucleus; LGN, Livingstone et al., 1991) have been reported as structural abnormalities in dyslexia (cf. Habib, 2000). Some of these structural differences (LGN and MGN) may also play a role in deviant neurocognitive functioning in dyslexics, both in the auditory and in the visual modality.

In a study comparing rhyming (subvocal rehearsal and storage in phonological working memory) or rhyming without necessary involvement of memory storage, Paulesu et al. (1996) found that both conditions activate a large perisylvian area (including Broca's and Wernicke's area), while the memory condition specifically also activated parietal operculum areas in normal controls. Dyslexics, however, showed activation only in the posterior part (inferior parietal cortex) in the memory task, whereas the rhyming task elicited activation only in the anterior part of the cortex (Broca's area). The common finding in both tasks was the absence of activation of the insular cortex, leading the authors to interpret the dyslexic deficits in terms of disconnection between anterior and posterior zones of cortical language areas.

In a visual experiment, using two visual conditions that differentially activated the magnocellular and parvocellular systems, Eden, VanMeter, Rumsey, Maisog, Woods et al. (1996) showed bilateral activation of the visual area for motion detection (MT/V5) after presentation of a moving-dots task in control subjects only, the dyslexics failed to show activity in this area. This finding was replicated by Demb, Boynton, and Heeger (1997) who were also able to show that this (magnocellular) processing deficit related to reading speed.

In a PET (positron emission tomography) study comparing Italian, English and French dyslexics, Paulesu and colleagues (2001) found that although the Italian dyslexics (who use a shallow orthography) performed better on reading tasks than the English and French dyslexics, all dyslexics were equally impaired on reading and phonological tasks relative to their controls. PET scans revealed that all dyslexics showed evidence of reduced activity during reading in a region of the left hemisphere, with the maximum peak in the middle temporal gyrus and additional peaks in the inferior and superior temporal gyri and middle occipital gyrus. These authors conclude that there is a universal neurocognitive basis for dyslexia and that differences in reading performance among dyslexics of different countries are due to different orthographies. This may, however, only hold for alphabetic writing systems, but not for logographic systems like Chinese. Reading impairment in Chinese is manifested by two deficits: one relating to the conversion of graphic form (orthography) to syllable, and the other concerning orthography-to-semantics mapping. Both of these processes are critically mediated by the left middle frontal gyrus and recently, Siok, Niu, Jin, Perfetti, and Tan (2008) showed that functional disruptions in this area are associated with impaired reading of the Chinese language.

Shaywitz et al. (2001) report an association between dyslexia and atypical cortical features in the left posterior parieto-temporal region and the left posterior occipito-temporal region, possibly causing impairment in processing the sound elements of language. On the basis of this finding, specific treatment was designed, which helped young dyslexics to develop neural circuitry in posterior portions of the left hemisphere sufficiently to enable them to read accurately and fluently (Shaywitz et al., 2001), thus showing plasticity of these neural circuits.

In a similar vein, Simos et al. (2002) employed phonology based remedial instructions in a group of dyslexic subjects and controls. Before the intervention, dyslexics displayed insufficient activation of the posterior part of the left superior temporal gyrus (STGp); an area involved in phonologic processing, but instead showed increased activity at the right hemisphere. After the treatment, their neurophysiological activation pattern appeared to be normalized, showing less activation in the right hemisphere, but increased activity in STGp, concurrent with improvement in reading skills.

For obvious reasons, remediation would benefit from early detection of dyslexia and for that reason prospective studies have been conducted. In such a study, Molfese (2000) recorded ERPs in response to speech and non-speech syllables from newborns and found that the results discriminated between newborn infants who 8 years later would be characterized as dyslexic, poor, or normal readers. Similarly, Guttorm et al. (2005) found that newborn ERPs to speech sounds can differentiate children with and without risk for dyslexia and that they are predictive of later language development and reading acquisition. Recently, van Leeuwen et al. (2008) reported anomalous auditory electrocortical mismatch responses elicited by differences between phonemes in 2-month-old infants at risk for dyslexia.

6.1.3 Future Directions

Cognitive neuroscience has already made important contributions to the understanding of the neurological substrate and the cognitive processes involved in dyslexia, which has already lead to the creation of interventions that show promising results.

However, more work remains to be done. Collaborative efforts between educational science and cognitive neuroscience can aim to resolve the debates between, for example, the dual route and connectionist models of word recognition which could, in turn, enable the creation of specific interventions.

Another relevant issue which is provided by cognitive neuroscience has to do with the promising progress in the area of early detection, as detailed above. Although current efforts illustrate progress in detection in the auditory modality, future efforts should also include the visual modality. These efforts once expanded, could aim toward establishing early detection protocols. Likewise, major contributions can come from research into the neurocognitive strategies used by children with a reading problem. There are indications that children develop other strategies to cope with the deficient linguistic-visual processing and that these strategies are not always the most efficient form of compensation. Functional brain imaging potentially provides a potent tool to evaluate the efficiency of the reading process and the automatisation of an overlearned response. Reading is not natural and the brain has to cope with different writing systems (Siok, Perfetti, Jin, & Tan, 2004), so the underpinnings of the (plasticity of the) process of reading acquisition (Maurer et al., 2007) should be further clarified. Based upon such neuroimaging studies, the learning process may be adapted in order to be optimally effective for the learner. Research programs into cognitive compensation and reading strategies involving auditory, visual, haptic sensory inputs and various strategies and types of learning materials should be devised and executed. Research should also further clarify the role of noise (and how it may possibly relate to learning in classroom situations) in both the visual and auditory modalities.

6.2 Dyscalculia

Although learning difficulties are just as common in mathematics as they are in reading, considerably less research has been done on mathematical learning difficulties (e.g., Rousselle & Noel, 2007; WHO, 1992) both with respect to the underlying causes with regard to the best educational practices. The term 'dyscalculia' is sometimes used to describe mathematical problems. Neuropsychologically, it is important to discern problems with the development of skills related to calculation and simple arithmetic from the focal neurocognitive deficits which have been described in terms of 'dyscalculia' in the neuropsychological clinic. Mathematical problems were identified later and have not been as well researched as dyslexia and are, therefore, less well understood.

A review of the literature reveals a variety of terms used to describe learning difficulties in the area of mathematics, including mathematical disabilities, mathematics learning difficulties, mathematics learning problems, mathematics learning disorders, mathematics learning disability, and mathematics learning deficiency, among others (Stock, Desoete, & Roeyers, 2006). These terms have often been defined by criteria such as falling below a given percentile on a standardized test of mathematics (Stock et al., 2006), with most authors using a cut-off point between the 25th and 35th percentile. The American Psychiatric Association defines dyscalculia as difficulty in learning arithmetic and failure to achieve adequate proficiency in arithmetic despite normal intelligence, scholastic opportunity, emotional stability, and necessary motivation (APA, 1994). Some authors include a specific reference to neurological deficits in their definition of dyscalculia. For example, Geary and Hoard (2001) define dyscalculia as deficits in the processing of numerical and arithmetical information associated with overt brain injury or presumed neurodevelopmental abnormalities. Mathematical learning problems are estimated to affect as little as 1% to as much as 7% of the school-age population (APA, 1994; Geary & Hoard, 2001), with most authors estimating close to 5%. It should be noted that mathematical learning difficulties that are defined by researchers as performance below the 25th and 35th percentile must necessarily include a larger group of children than the approximately 5% with dyscalculia. Again, the reader should bear in mind that the term 'dyscalculia' is used differently by different professionals or research disciplines with the most strict definition stating that dyscalculia should only be used in case of actual or anticipated brain dysfunction, whereas others use the term in a psychometric sense.

6.2.1 Education

Although more research is needed to investigate the cognitive characteristics of children with mathematical learning difficulties, children with mathematical learning difficulties who are good readers have been shown to have deficits in the ability to retrieve the answers to simple arithmetic problems, such as 5 + 3, from long term memory, a skill referred to as math fact fluency, whereas children with both mathematical and reading difficulties have not only deficits in math fact fluency, but also in problem solving skills (Geary, Hamson, & Hoard, 2000).

The signs and symptoms of mathematical problems can vary greatly (Geary & Hoard, 2001; Shalev & Gross-Tsur, 2001). Some (younger) children with mathematical problems have difficulties with number sense, the early understandings of numerical quantities and their relations. Many children with mathematical problems may demonstrate difficulty learning number facts (e.g., 2 + 5, 3 × 6) and with the retrieval of such facts from memory. Apparently as a result of these retrieval deficits, children with mathematical problems tend to use inefficient strategies that place a greater demand on, for example, working memory. Children with mathematical problems may make errors resulting from incomplete procedural knowledge necessary for complex problems in addition, subtraction, multiplication, or division. These children may also confuse arithmetic symbols (e.g., +, −, ×, ÷) and make procedural errors as a result. The diagnosis of mathematical problems is based on assessment of the child's arithmetic skills and can best be determined by a discrepancy between the intellectual potential of the child and his or her arithmetic achievement (Shalev & Gross-Tsur, 2001). This presupposes the existence of reliable standardized tests that measure all (age appropriate) aspects of numeracy and mathematics.

6.2.2 Cognitive Neuroscience

Much cognitive neuroscience research has been done in recent years with the goal of revealing the neural substrate of numerical cognition and a short review of this research can be found in Section 5.2 of this volume. Evidence suggests that areas in the (bilateral) parietal cortex of the brain, including areas involved in verbal and visuospatial processing, are involved in number processing (Dehaene et al., 2003). Interestingly, Castelli et al. (Castelli, Glaser, & Butterworth, 2006) recently showed that counting (particularly affected in dyscalculia) and estimating have different neural substrates in the intraparietal sulcus. Cognitive processes such as executive function and working memory have also been shown to be important for number processing, both of which rely on frontal and parietal areas (Collette et al., 2006; D'Esposito et al., 1999). Neuropsychological tests can be used to help determine the specific cognitive deficits underlying mathematical problems in a particular child. Consequently, it has been suggested that remedial education for children with mathematical problems should employ interventions appropriate for the underlying neuropsychological deficits of the particular child, for example, perceptual and visuospatial or verbal and auditory-perceptual (Rourke & Conway, 1997). There is even evidence that the use of targeted interventions can lead to changes in the brain itself. Learning new number facts or processes alters brain activity (Delazer et al., 2003; Ischebeck et al., 2006). Recent advances in neuropsychological research show that there is a developmental factor involved. Some children might be later than other children in the development of elementary functions which are important for calculation and arithmetical operations. Estimation, mental rotation and spatial processes related to number sense seem to be important. Clinical neuropsychological research into the implications of focal brain damage also underscores that mathematical abilities require proficiency in quite a number of different neurocognitive functions. More research should be done in order to link these findings to those done in cognitive psychological research, cognitive neuroscience and educational research.

6.2.3 Future Directions

Although much has been learned about the neural substrate responsible for numerical cognition in recent years, much work remains to be done. In particular, more work is required to examine the precise neurocognitive underpinnings of dyscalculia in the broader sense of mathematical problems, as well as the strict sense. An understanding of these cognitive underpinnings can be used to design neuropsychological tests that can be used to more reliably diagnose dyscalculia and to suggest specific areas to target in interventions. Of major relevance in this respect are the developmental patterns and the complex nature of the skill of mathematical operations, which require the identification of cognitive subprocesses and their interaction and developmental profile. The development and functional significance of the differentiation between 'many' and 'much' in the intraparietal sulcus (Castelli, Glaser, & Butterworth, 2006), as a candidate marker for dyscalculia, needs to be clarified. The results of studies that show changes in the brain in response to learning and interventions (Delazer et al., 2003; Ischebeck et al., 2006) are exciting and more of such work needs to be done. Paradigms from cognitive neuroscience offer new ways to judge the effectiveness of one intervention in comparison to another.

Chapter 7
Issues from Neuroscience

The present volume takes educational issues as a starting point and looks at possible contributions which could be given from the point of view of cognitive neuroscience. Issues that directly arise from the neurosciences which might be of relevance for education are only described in relation to educational issues. The interested reader is referred to recent reviews and 'opinion' articles which take a more cognitive neuroscience stance (e.g., Ansari & Coch, 2006; Casey, Getz, & Galvan, 2008; Jolles, 2007a,b; OECD, 2002, 2007; Steinberg, 2008) and other papers, mentioned in the Introduction. Yet, two issues deserve a short elaboration, because they are pertinent to the scientific findings and directions described in earlier sections. This concerns the issues of 'plasticity' and 'maturation'.

7.1 Plasticity

The term 'plasticity' refers to the capacity of nervous tissue to change in structure and function in response to factors which can be described as 'environmental'. These factors act on the level of the neuron or its substructures such as dendrites and dendritic spines but also – on a more macroscopic level – at the level of aggregates of neurones (e.g., neuronal networks). A plastic reorganization of nervous tissue occurs in relation to neurotransmitters and neurohormones and other chemical entities such as toxic factors. In addition to that, learning and psychological processes as well as emotional experiences can change brain structure and function due to the fact that 'plasticity' is an inherent property of nervous tissue. This change is at the structural level, which means that the shape and structure of the neuron and neuronal assemblies – and thus the brain – changes (Kolb & Whishaw, 2008). Because of the inherent capacity of the brain to adapt to a changing environment, plasticity lies at the basis of brain development and maturation, and thus provides the mechanisms which underlie adaptation and learning. There are three aspects of plasticity which should be described in more detail to serve as the basis for the notions described in earlier sections.

First, nervous tissue – and thus the brain – remains plastic up till very high age. This notion comes from decades of research performed with both animals and humans (e.g., Buonomano & Merzenich, 1998; Weinberger, 1995). This underscores

T. de Jong et al., *Explorations in Learning and the Brain*,
DOI 10.1007/978-0-387-89512-3_7,

the notion that learning is possible up till very old age. There is now a wealth of information in favour of the notion that an active lifestyle with cognitive activities, social and/or physical activities can protect against brain and cognitive deterioration (Bosma et al., 2002; Karp et al., 2006; Wilson et al., 2002). These important findings support approaches towards the concept of 'Lifelong Learning'.

Second, plastic changes in nervous tissue on both the neuronal level and the level of neuronal networks make it possible for a patient to recover. In addition, the patient is able to compensate in case of brain damage or malfunctioning neural network due to infarcts, brain trauma, toxic factors, and other influences. This means that when brain areas are damaged, for example after a stroke, either the brain cells themselves recover, or there are other brain cells or aggregates of cells which take over the functions of the damaged part (Kolb & Whishaw, 2008; Robertson & Murre, 1999). In relation to this, other neuronal networks develop, and the brain is able to compensate for the functions which are lost. Besides age, practice is an important factor in recovery or compensation. Indications exist that the brain can be stimulated by dedicated forms of practice/training and thereby recover in the most effective way (Cicerone et al., 2005; Taub et al., 2006).

A third aspect of plasticity is relevant in relation to 'normal' learning of motor skills and even cognitive knowledge. Such findings have been reported in recent years in adults who had practiced a particular skill and who were subjected to brain imaging analysis by MRI. Interesting findings in this respect were those reported by Draganski et al. (2004). They taught 12 people how to juggle, and compared structural MRI scans of their brains before and after they mastered juggling. It was shown that grey matter volume increased in the bilateral mid temporal area and in the left posterior intraparietal sulcus. This change was not seen in 12 controls who did not learn to juggle. This implies that plasticity not only plays a role in development, or after brain damage, but also in learning a particular skill. Others have found changes in the hippocampal area (a region known to be involved in learning and in memory for places) in taxi drivers in the city of London (Maguire et al., 2000). Their findings can be interpreted in terms of plastic adaptations of specialized brain structures towards environmental demands. Many recent examples can be given. In general, these findings imply that 'learning' as observed in educational settings is always accompanied by changes in nervous tissue and in the effectiveness of information processing in the brain. As children, youth and also adults adapt to a constantly changing environment, this means that peers, parents, teachers and others – psychosocial context- have a major role in shaping the efficiency of brain plasticity and thus adaptation.

7.2 Maturation

Maturation of the brain in relation to cognitive and emotional development has become a major topic in neuroscience. It has been acknowledged in recent years that particular parts of the brain and their connections develop in middle and late adolescence and are not fully matured till well in the third decade of life. The medial prefrontal cortex and other regions in the frontal lobes, but also tertiary areas in the parietal lobe are among them. Particular brain areas may be fully developed around birth whereas others become mature in early childhood, in late childhood or in early, middle or late adolescence. This development continues probably to around 25 years of age (Giedd, 2004; Gogtay et al., 2004). Brains of boys, generally, mature at a slower pace and are fully developed some years later than those of girls (Giedd, 2008). There is individual variability in this process and it is very probable that the nature of the psychosocial environment – family, peers – and the learning context – both psychosocial and school factors – might have a modulating role and even 'guide' this maturation. These new findings originate from research with structural and functional brain scans in a developmental perspective, published in the past five to seven years (Blakemore & Frith, 2005; Giedd, 2004; Gogtay et al., 2004; Paus, 2005; Steinberg, 2005).

The brain areas which develop relatively 'late' – in middle and late adolescence – correspond to particular aspects of executive functioning. Executive functions are concerned with planning, executive control, concept shifting, efficient processing in working memory, attentional processes and the so-called 'self-evaluation' and 'social monitoring' (see also Section 2.6) (Blakemore & Choudhury, 2006). It is important to acknowledge that the start of this development is guided by biological factors in which the genes play a role. In recent years, it has been found that 'epigenetic' factors are responsible for the expression of genes, leading to the production of proteins. These proteins guide the development of nervous tissue by serving as structural elements of the cell, as receptors, enzymes, or that they guide the cell machinery by other means (for interested readers: further information can be found in general handbooks). Although the development of the various brain regions and their connections are guided by the genetic 'blueprint' and other biological factors, environmental factors are responsible for the development within the constraints given by genes and biology. The brain is plastic, which means that the proper growth, the proper development of brain networks and the pruning of connections and optimization of interneuronal communication is guided by *behaviour, sensory and motivational information* (Kolb & Whishaw, 2008). Thus, environmental factors determine the proper functioning of these brain structures and thus higher cognitive, psychological and social functioning. In other words, it is not the case that this development is an autonomous biological process. It is the other way around: the brain has to be nurtured by social and cognitive stimuli and by psychomotor actions in order to pass the final maturation over three phases in adolescence, until 22–25 years of age. The structures in the prefrontal cortex (e.g., anterior cingulate area) are thought to have strong connections to the limbic areas, and areas in the hypothalamus and thalamus. Their role is to enable an efficient behavioural planning in keeping with the motivational and emotional processes. Self-evaluation and social monitoring are thought to be guided by these structures. This makes the medial prefrontal cortex and its connections to other prefrontal, limbic and subcortical areas of importance in relation to (educational) learning and the role of motivation and emotion.

The results of neuroscientific research imply that learners (child, adolescent, youth and young adults even up to 22–25 years of age) need support, especially in tasks that are dependent upon various executive functions. This support can be cognitive or psychosocial in nature (parents, teachers, peers and others) or come from properly designed educational material (including software). Neuroscientific research suggests that this support is not only needed to properly perform such self-regulated learning tasks but that by performing them in a properly scaffolded setting the development of the brain is also fostered.

Chapter 8
Conclusion

This volume has sought to identify promising areas of research in which educational research and cognitive neuroscience could come together. It is clear that we are at the edge of an exciting new field of research. This has also been stated in recent papers in the neuroscience and cognitive science field (e.g., Ansari & Coch, 2006) in their paper on the need to build multiple bridges between cognitive neuroscience and education). At this very moment, the new domain is promising but there are as yet not many findings which have direct consequences for educational practice, though some of them may have consequences of educational research. There are some recommendations for implementation of neuroscience findings in education which have a more 'general' character, other studies are at a level of detail that abstractions to educational research (and certainly practice) still need to be made; yet, there are not many findings which can directly be translated into educational design. As early as 1991, Caine and Caine (1991) presented 12 recommendations for education based on neuroscientific research. These recommendations include statements such as: all learning is physiological, the search for meaning is innate; the search for meaning occurs through patterning; emotions are critical to patterning, learning is developmental etc. Current recommendations often equal Caine and Caine's recommendations in terms of generality and a lacking overall view. Such a general approach may easily lead to the use of what are called 'neuromyths' (OECD, 2007). The other side of the coin is that quite a few studies from neuroscience focus on cognitive functions that are at a level of detail that is fine-grained compared to processes from educational theories. For example, to present findings on the phenomenon of insight we have assumed that drawing relations between semantically distant concepts is involved (see Section 2.4.2) and for regulative skills we have turned to processes such as error detection (see Section 2.6.2). Defining research that renders results at the right level of abstraction will be one of the challenges for the field.

Another reason why conclusions from cognitive neuroscience to educational research and/or practice are not easily made lays in the fact that the learning process is diverse and involves a vast domain of different applications, varying from knowledge learning, via learning psychomotor acts to learning social-emotional skills and higher cognitive processes including self-regulation and self-initiated learning. In

T. de Jong et al., *Explorations in Learning and the Brain*,
DOI 10.1007/978-0-387-89512-3_8, © Springer Science+Business Media, LLC 2009

addition, many factors are known to be of major importance for learning, including instruction-related factors, child/learner-related factors (including age, sex, and biological factors) and context related factors (social class, parental education, culture). Accordingly, there is not a simple step from cognitive neuroscience research into the educational setting. Moreover, recommendations are made for different areas of education without considering integration into a coherent curricular approach.

The time has come to conduct new types of research that will provide us with adequately detailed and applicable guidelines for educational design based on neuroscientific data. As was indicated in the present volume, neuroscience research may prove to be of critical relevance for educational theories or areas of research. In this volume, we have, partly based on the expert meeting that was held, identified themes which elaborate on major routes described earlier by Jolles et al. (2005), notably those which are most relevant for further development of educational research. Thus, the present volume elaborates upon: (a) multimedia learning (Mayer, 2001) for which findings regarding learning from multiple representations and multimodal processing could be relevant; (b) cognitive load (Sweller et al., 1998), for which findings on neurological correlates of cognitive load and attention are of interest; (c) problem solving (Ohlsson, 1992), for which, for example, indicators for insight are of relevance; (d) implicit learning (Reber, 1989) that is (partly) associated with activation in different brain regions than explicit learning); (e) metacognitive and regulative skills (Flavell, 1971) for which the neuroscientific processes of conflict resolution, error detection, causal thinking, and planning are of relevance; (f) social-observational learning (Bandura, 1986) and social-emotional learning for which the research on the mirror-neuron system seems important; (g) affective processes in learning (Boekaerts, 2003) for which students' emotional reactions to learning material can be charted; (h) language acquisition and literacy development, the cognitive and brain processes involved in learning a foreign language, and the implications of exposure to multiple languages at an early age; (i) numeracy and mathematics learning, including work on mathematics learning difficulties (Rousselle & Noel, 2007) could profit from neuroscience research efforts to locate specific mathematical processes (e.g., number processing and semantic activities) and the involvement of executive processes; and (j) learning disabilities (Lerner & Kline, 2006) and severe learning problems, such as dyslexia and dyscalculia, for which neuroscientific methods for early detection and the effects of intervention are central. It also addresses two issues in neuroscience (plasticity and maturation) that can have consequences for education.

Depending on the nature of the findings which have been collected in preceding years, and will be gathered in the near future, several interpretative steps are required to identify what interesting interfaces for interdisciplinary research could be, or what findings from neuroscience in these areas could contribute to educational research. Examples of pertinent questions include: 'does this provide implications for designing instruction, that is, to shape and support learning?', 'does this deepen our insight into neurocognitive processes and skills involved in self-initiated learning?', 'does this provide mechanisms to understand the efficient development of elementary skills and the subsequent application in a more complex educational performance?'. Thus, findings from neuroscience in terms of activation patterns or neural changes show that types of learning (tasks) are correlated with activation or growth of specific brain areas. Although this is highly informative, additional interpretation is necessary to link brain area activation or growth to cognitive processes (e.g., Henson, 2006; Poldrack, 2006). In our Amsterdam expert workshop it became clear that, although in principle neurological indicators can be identified for many of the cognitive processes that are relevant in educational settings, the link between the neurological indicator (or activated brain area) and cognitive function is not always straightforward. Also in this volume we have seen that for the same cognitive construct (e.g., working memory) different indicators are used and it is not always clear what the 'best' neurological indicator should be, or, alternatively, if the cognitive construct itself should better be reconsidered. The latter would mean that educational theories revise their constructs on the basis of neuroscientific findings.

A further step is to translate neuroscientific findings into practical considerations for use in the classroom. Besides cognitive processes, also interactional skills, motivational processes, social and emotional monitoring and self-evaluation of the learner (to name but a few) are needed. Based on the work by Byrnes and Fox (1998) we can identify two directions to interpret these types of results. First, that research in cognitive neuroscience (including social and affective neuroscience) can aid educational insights as to the nature of cognitive processes while students are engaged in learning tasks, and secondly, cognitive neuroscience may aid educational researchers in their search to resolve conflicts in existing educational theories. In addition, findings from neuroscience research also involve behavioural measures or measures of learning outcomes. These measures might confirm or corroborate findings from educational research, thereby strengthening educational theories with knowledge of underlying cognitive and brain mechanisms of observed effects on learning as well as cognitive neuropsychological insights into learning and educational performance of individual learners, given their developmental stage, psychosocial context, biopsychological variables and other aspects. It should be noted that all these emerging relations bloom up between neuroscience and the most 'cognitive' area of educational science, namely educational psychology. If we come to the broader educational issues that have to do, for example, with classroom organization (e.g., the optimal number size of a class and school dropout), the bridge between neuroscience and educational science is even larger.

To bring the complex fields of educational and neuroscientific research together we also need to bridge the methodological approaches used in both scientific fields. It should also be borne in mind that the fields as such are multidimensional in themselves with researchers focussing on instruction, on knowledge transfer, on attentional, motivational, or psychological processes in individual learners or on various aspects of educational performance and/or age or intellectual level. One interesting aspect concerns the granularity of research. At this point, tasks used in neuroscience are often short, decontextualized, and isolated, whereas in educational research tasks are often long (ranging from one lesson to a series of lessons), content rich and diverse, and embedded in a complex (social) environment (the classroom, trainee post, at home). This not only hampers the translation of results from neuroscientific research into educational practice, but also calls for new methodological approaches that will bridge the gap between the two scientific approaches. Part of creating this bridge is that neuroscientific data collection techniques (EEG, PET, fMRI) should be made applicable to tasks and situations as they typically appear in educational research, in which for example complex tasks are used over a prolonged period of time in which users are allowed to move their heads freely. With rapid technological developments, however, this may be possible in the near future. For example, wireless EEG equipment integrated in caps is available that provides freedom of movement and can therefore be used in real-world tasks (Berka et al., 2008). Fugelsang and Dunbar (2005) have provided an example of how cognitive neuroscience can incorporate more complex and educationally relevant tasks. The same applies to approaches such as proposed and used by Blakemore, Den Ouden, Choudhury, and Frith (2007).

The present volume may provide some routes to follow in the search for potent paradigms and good scientific models which can guide a science-based educational innovation which our society calls for. We think it reflects some of the most important trends that can be observed in the literature, whereas it does not pretend to provide a complete coverage of the domains or to give an in-depth evaluation of all relevant issues. This text will primarily act as a starting point for creating an agenda for educational science research that incorporates neuroscientific theories and techniques.

Appendix A
List of Participants in the Amsterdam Workshop

Jos Beishuizen	VU University Amsterdam – Centre for Educational Training, Assessment and Research	The Netherlands
Monique Boekaerts	Leiden University – Faculty of Social and Behavioural Sciences	The Netherlands
Annemarie Boschloo	Maastricht University – Department of Neuropsychology	The Netherlands
Brian Butterworth	University College London – Institute of Cognitive Neuroscience & Department Psychology	United Kingdom
Eveline Crone	Leiden University – Leiden Institute for Brain and Cognition (LIBC)	The Netherlands
Tamara van Gog	Open University Netherlands – Educational Technology Expertise Centre	The Netherlands
Peter Hagoort	Radboud University Nijmegen Donders Institute for Brain, Cognition, and Behaviour: Centre for Cognitive Neuroimaging	The Netherlands
Janet van Hell	Radboud University Nijmegen – Faculty of Social Sciences	The Netherlands
Katrin Hille	Transferzentrum für Neurowissenschaften und Lernen	Germany
Bernadette van Hout-Wolters	University of Amsterdam – Graduate School of Teaching and Learning (ILO)	The Netherlands
Paul Howard-Jones	University of Bristol – Graduate School of Education	United Kingdom
Kathleen Jenks	Radboud University Nijmegen – Behavioural Science Institute	The Netherlands
Jelle Jolles	Maastricht University – Department of Neuropsychology/Institute of Brain and Behavior	The Netherlands
Ton de Jong	University of Twente – Faculty of Behavioral Sciences	The Netherlands
Theo van Leeuwen	University of Twente – Faculty of Behavioral Sciences	The Netherlands

Jeroen van Merriënboer	Open University Netherlands – Netherlands Laboratory for Lifelong Learning	The Netherlands
Alexander Renkl	University of Freiburg, Psychological Institute, Educational and Developmental Psychology	Germany
Todd Rose	Harvard University – Graduate School of Education/Center for Astrophysics	United States of America
Ralph Schumacher	ETH Zürich, Institut für Verhaltenswissenschaften	Switzerland
Stephan Schwan	Knowledge Media Research Center – KMRC	Germany
Bert de Smedt	K.U. Leuven – Department of Educational Sciences, CIP&T	Belgium
Elsbeth Stern	ETH Zürich, Institut für Verhaltenswissenschaften	Switzerland
Hanna Swaab	Leiden University – Faculty of Social and Behavioural Sciences	The Netherlands
Natasha Tokowicz	University of Pittsburgh – Learning Research and Development Center	United States of America
Ludo Verhoeven	Radboud University Nijmegen – Faculty of Social Sciences	The Netherlands
Lieven Verschaffel	K.U. Leuven – Department of Educational Sciences, CIP&T	Belgium

Appendix B
Executive Summary

We might safely assume that teachers and educational researchers would love to have the chance to open students' scalps and look directly to what is going on there. How nice would it be to directly identify impasses in problem solving behaviour, insight, the effort students exert, their level of motivation, use of learning material etc.? Educational research and educational practice would profit a lot from those types of information. Cognitive neuroscience opens the door towards such developments, but at the moment, one has to be modest with respect to the expectations. Recommendations which come from cognitive neuroscience for implementation into the field of education are presently either formulated on a very 'general' level, or are so fine grained that the relation with educational research (let alone practice) is not clear. The insights from cognitive neuroscience definitely bears promise, but quite a lot of fundamental and applied research has yet to be performed before the results can be directly applied in educational settings. The present volume has set out to define a research agenda by identifying actual themes in educational research for which neuroscientific data are of relevance either by providing further support for already known phenomena or by providing us with insight into phenomena that remained uncovered until now since behavioural techniques are insufficient to identify those phenomena.

Thus, the present volume elaborates upon: (a) multimedia learning, for which findings regarding learning from multiple representations and multimodal processing could be relevant; (b) cognitive load, for which findings on neurological correlates of cognitive load and attention are of interest; (c) problem solving, for which, for example, indicators for insight are of relevance; (d) implicit learning that is (partly) associated with activation in different brain regions than explicit learning; (e) metacognitive and regulative skills for which the neuroscientific processes of conflict resolution, error detection, causal thinking, and planning are of relevance; (f) social-observational learning and social-emotional learning for which the research on the mirror-neuron system and research in the domain of social cognitive neuroscience seems important; (g) affective processes in learning for which students' emotional reactions to learning material can be charted; (h) language acquisition and literacy development, the cognitive and brain processes involved in learning a foreign language, and the implications of exposure to multiple languages at an early age; (i) numeracy and mathematics learning, including work on mathematics learning difficulties could profit from neuroscience research efforts to locate specific mathematical processes (e.g., number processing and semantic activities) and the involvement of executive processes; and (j) learning disabilities, and severe learning problems, such as dyslexia and dyscalculia, for which neuroscientific methods for early detection and the effects of intervention are central.

For all of these themes we have identified developments in neuroscientific research that are of direct relevance. To bring the complex fields of educational and neuroscientific research together we would, however, also need to bridge the methodological approaches as used in both scientific fields. Part of creating this bridge necessitates making neuroscientific data collection techniques (EEG, PET, fMRI) more applicable to tasks and situations as they typically appear in educational research, in which for example complex tasks are used over a prolonged period of time. With rapid technological developments, however, part of these problems may be solved in the near future. Wireless EEG equipment, for example, integrated in caps has become available, providing freedom of movement.

Though many (theoretical, practical, ethical) issues are still to be overcome, the present text may provide routes to follow in the search for potent paradigms and good scientific models which can guide a science-based educational innovation which our society calls for.

Appendix C
About the Authors

Ton de Jong studied cognitive psychology (cum laude) at the University of Amsterdam and received a PhD in Technological Sciences from the Eindhoven University of Technology on the topic 'problem solving and knowledge representation in physics for novice students'. Currently he is full professor of Educational Psychology at the University of Twente, Faculty of Behavioral Sciences where he is department head of the department Instructional Technology. In 2001/2002 he has also been (part-time) full professor at the Institute for Knowledge Media at the University of Tübingen (Germany. His main interests are in problem solving in science, inquiry (computer-simulation based) learning environments, learners' cognitive processes, instructional design, and man-machine interfaces. He was project manager of several EU and NWO funded projects He has been scientific director of the Dutch national school for educational research (ICO) from 2003–2008. IP SCY (Science Created by You) that will develop a multimedia learning environment for science topics. (Ton de Jong, University of Twente, Faculty of Behavioral Sciences, PO BOX 217, 7500 AE Enschede, The Netherlands. Email a.j.m.dejong@utwente.nl)

Tamara van Gog holds a Master's in Educational Psychology from Tilburg University (2001), and received her PhD in Educational Technology from the Open University of The Netherlands (OUNL) in 2006. She is currently Associate Professor at the Centre for Learning Sciences and Technologies of the OUNL. From November 2007 until June 2008, she was a visiting researcher at the Knowledge Media Research Center in Tübingen, Germany, supported by a Rubicon grant from the Netherlands Organization for Scientific Research. Her own research and that of her PhD students focuses on example-based learning, learning from animations, expertise, (self-)assessment, and uncovering cognitive processes using techniques such as verbal protocols and eye tracking. (Tamara van Gog, Open University of the Netherlands, Centre for Learning Sciences and Technologies and Netherlands Laboratory for Lifelong Learning, PO BOX 2960, 6401 DL Heerlen, The Netherlands. Email: Tamara.vanGog@ou.nl).

Kathleen Jenks holds a Bachelor's degree in Clinical Psychology from the University of Wisconsin Oshkosh, a Masters in Neuropsychology from Radboud University Nijmegen is about to complete her PhD at Radboud University Nijmegen,

The topic of her PhD project is the development of arithmetic ability in children with cerebral palsy. Her current research interests include working memory, executive function and numerical cognition. (Radboud University Nijmegen, Behavioural Science Institute, Montessorilaan 3, 6525 HE Nijmegen, the Netherlands, k.jenks@pwo.ru.nl or kjenks@hotmail.com).

Sarah Manlove Sarah Manlove recently completed her PhD. in instructional design at the University of Twente. Her focus of research was on the design of support for self-regulatory processes for students involved in an inquiry learning task in a technology-enhanced learning environment. Prior to attaining her PhD. Sarah attended Penn State University for her Masters in Instructional systems after working in diverse educational settings such as in industrial training and design, elementary education and English as a Second language instruction. (email: smanlove@hughes.net)

Janet van Hell holds an MA degree in experimental psychology from the University of Amsterdam (cum laude) and received her Ph.D. from the same university in 1998. She is currently associate professor at the Radboud University Nijmegen, the Netherlands, and visiting professor of Psychology at the Pennsylvania State University, USA. Her research focuses on second language learning and bilingualism as well as later language development in children with typical or atypical language development. Her research has an interdisciplinary orientation, and she and her students combine behavioral, neuropsychological, and linguistic techniques to study language development and language processing. Her research is supported by grants from, amongst others, the Dutch National Science Foundation, the Dutch Ministry of Education, Culture, and Science, National Science Foundation (NSF), and research institutes within the Radboud University Nijmegen and Penn State University. Janet van Hell is member of the young investigators section of the Royal Netherlands Academy of Arts and Sciences. (Radboud University Nijmegen, Behavioural Science Institute, Montessorilaan 3, 6525 HE Nijmegen, the Netherlands; the Pennsylvania State University, Department of Psychology, 626 Moore Building, University Park, PA 16802 USA. Email j.vanhell@pwo.ru.nl or jgv3@psu.edu)

Jelle Jolles is neuroscientist and educational neuropsychologist. He is director of the Brain & Learning Centre Amsterdam and the Amsterdam Institute for Research in Education/AZIRE. His research has its focus upon normal, succesful and pathological conditions of cognitive development and aging. Over the past 25 years, he and his PhD students have investigated the interaction of biological and psychosocial factors in their effect upon cognitive functioning, notably learning and memory. His centre is involved in fundamental research but also in applied research and in the implementation of research insights into applied settings (education and health care). There are several foci in the research of the centre. Research line 1 is upon children and adolescents and the mechanisms underlying cognitive development and learning performance. 'Talent development', and determinants of normal, succesful and pathological conditions of learning are at the heart of this research line. Research line 2 is upon 'Lifelong Learning' through adulthood and both normal and succesful aging. Research line 3 is dedicated to neuropsychiatric disorders in children, adolescents and adults, with a focus upon ADHD, learning disorders, cognitive disorders in adults, and Alzheimer and its prodromes. The research group makes use of large scale crosssectional-longitudinal studies (children, adolescents, adults) with neuropsychological methods, neuroimaging but also dedicated 'evidence based' interventions. Jolles has been chairman of the Dutch 'committee on Brain & Learning' which has been installed by the Dutch Science Council and the ministry of Education. He actively advocates a constructive dialogue between neuroscience, educational science, (developmental) psychology and practitioners with the aim to help develop educational reforms which are based upon scientific insights. (Jelle Jolles, Faculty of Psychology and Education\VU University Amsterdam, van der Boechorststraat 1, 1081 BT Amsterdam, Netherlands. Email: j.jolles@psy.vu.nl).

Jeroen van Merriënboer has a Masters degree from the Free University of Amsterdam in experimental psychology with a specialization in psychophysiology, and a PhD degree from the University of Twente in instructional technology (cum laude). Most of his research is related to cognitive load theory (together with John Sweller) and his four-component instructional design model (4C/ID-model), which builds instructional blueprints from four interrelated components: (1) learning tasks (e.g., worked-out examples, completion tasks, conventional problems etc.), (2) supportive information, which helps learners—learn to—perform the problem-solving and reasoning aspects of learning tasks, (3) procedural information, which helps learners—learn—to perform the routine aspects of learning tasks, and (4) part-task practice, which helps learners to develop selected routine aspects of learning tasks to a high level of automaticity. This model is taught in educational instructional-design programs around the world and broadly applied in educational practice. van Merriënboer is scientific director of the Interuniversity Center for Educational Research, which is a joint PhD program of 10 Dutch universities. In May 2009, he will move from the Open University of the Netherlands to Maastricht University, where he will hold a professorship in Learning and Instruction. (Jeroen van Merriënboer, Maastricht University, PO Box 616, 6200 MD Maastricht).

Theo van Leeuwen studied psychology (with a specialization in psychophysiology) at Utrecht University, the Netherlands, where he also received his PhD on studies into the effects of psychopharmacological substances (benzodiazepines) on sustained attention and event-related brain activity. Since then he was involved in a European multi-center study on the neurobiological aspects of Attention Deficit Hyperactivity Disorder (ADHD), at the department of Child and Adolescent Psychiatry at the University of Zürich, Switzerland, and in a nation-wide longitudinal prospective study aimed at finding a neurobiological marker for developmental dyslexia, at the University of Amsterdam, the Netherlands. Currently, he is at the department of Instructional Technology at Twente University, the Netherlands, to explore the field of neuroscience and education. (Theo van leeuwen, University of Twente, Faculty of Behavioral Sciences, PO BOX 217, 7500 AE Enschede, The Netherlands. Email t.h.vanleeuwen@utwente.nl)

Annemarie Boschloo is a PhD-student in educational neuropsychology at the Centre for Brain & Learning. Jelle Jolles is her promoter and she assists him on this project. Current topics of interest are executive functions in adolescence and sleep. (Annemarie Boschloo, VU University Amsterdam, Faculty of Psychology and Education, van der Boechorststraat 1, 1081 BT Amsterdam. Email: am.boschloo@psy.vu.nl)

References

Abutalebi, J., Stefano, F. C., & Perani, D. (2005). What can functional neuroimaging tell us about the bilingual brain? In J. F. Kroll, & A. M. B. de Groot (Eds.), *Handbook of bilingualism: Psycholinguistic approaches* (pp. 497–515). New York: Oxford University Press.

Adams, J. W., & Hitch, G. J. (1997). Working memory and children's mental addition. *Journal of Experimental Child Psychology, 67*, 21–38.

Ainsworth, S. E. (1999). The functions of multiple representations. *Computers and Education, 33*, 131–152.

Ainsworth, S. E., Bibby, P. A., & Wood, D. J. (1997). Information technology and multiple representations: new opportunities – new problems. *Journal of Information for Teacher Education, 6*, 93–104.

Aizenstein, H. J., Stenger, V. A., Cochran, J., Clark, K., Johnson, M., Nebes, R. D., et al. (2004). Regional brain activation during concurrent implicit and explicit sequence learning. *Cerebral Cortex, 14*, 199–208.

Amodio, D. M., & Frith, C. D. (2007). Meeting of minds: the medial frontal cortex and social cognition. *Nature Neuroscience, 7*, 259–277.

Ansari, D., & Coch, D. (2006). Bridge over troubled waters: education and cognitive neuroscience. *Trends in Cognitive Sciences, 10*, 146–151.

APA. (1994). *Diagnostic and statistical manual of mental disorders* (4th ed., Vol. IV). Washington, DC: American Psychiatric Association.

Bach-y-Rita, P., Danilov, Y., Tyler, M., & Grimm, R. J. (2005). Late human brain plasticity: Vestibular substitution with a tongue BrainPort human-machine interface. *Plastcidad y Restauración Neurológica, 4*, 31–34.

Backhaus, J., Born, J., Hoeckesfeld, R., Fokuhl, S., Hohagen, F., & Junghanns, K. (2007). Midlife decline in declarative memory consolidation is correlated with a decline in slow wave sleep. *Learning & Memory, 14*, 336–341.

Baddeley, A. D. (2000). The episodic buffer: A new component of working memory? *TRENDS in Cognitive Sciences, 4*, 417–423.

Baddeley, A. D., & Hitch, G. J. (1974). Working memory. In G. A. Bower (Ed.), *The psychology of learning and motivation: Advances in research and theory* (Vol. 8, pp. 47–89). New York: Academic Press.

Badgaiyan, R., & Posner, M. I. (1998). Mapping the cingulate cortex in response selection and monitoring. *Neuroimage, 7*, 255–260.

Baker, D. R., & Piburn, M. D. (1997). *Constructing science in middle and secondary school classrooms*. Needham Heights, MA: Allyn & Bacon.

Bandura, A. (1977). *Social learning theory*. Englewood Cliffs, NJ: Prentice Hall.

Bandura, A. (1986). *Social foundations of thought and action: A social cognitive theory.* Englewood Cliffs, NJ: Prentice Hall.

Beauchamp, M. S., Lee, K. E., Argall, B. D., & Martin, A. (2004). Integration of auditory and visual information about objects in superior temporal sulcus. *Neuron, 41*, 809–823.

Berka, C., Chung, G. K. W. K., Nagashima, S. O., Musacchia, A., Davis, G., Johnson, R., et al. (2008). *Using interactive neuro-educational technology to increase the pace and efficiency of rifle marksmanship training*. Paper presented at the Annual Meeting of the American Educational Research Association.

Bernhardt, E. B. (1991). *Reading development in a second language: Theoretical, empirical, and classroom perspectives*. Norwood, NJ: Ablex.

Bernhardt, E. B. (2000). Second-language reading as a case study of reading scholarship in the 20th century. In M. L. Kamil, P. B. Mosenthal, P. D. Pearson, & R. Barr (Eds.), *Handbook of reading research* (Vol. 3). London: Lawrence Erlbaum Associates.

Berninger, V. W., & Richards, T. (2002). *Brain literacy for educators and psychologists*. San Diego: Academic Press.

Blakemore, S. J., & Choudhury, S. (2006). Development of the adolescent brain: implications for executive function and social cognition. *Journal of Child Psychology and Psychiatry, 47*, 296–312.

Blakemore, S. J., Den Ouden, H., Choudhury, S., & Frith, C. (2007). Adolescent development of the neural circuitry for thinking about intentions. *SCAN, 2*, 130–139.

Blakemore, S. J., & Frith, U. (2005). *The learning brain: Lessons for education*. Oxford: Blackwell Publishing.

Boekaerts, M. (2003). Toward a model that integrates motivation, affect, and learning. *British Journal of Educational Psychology Monograph Series, II*, 173–189.

Boekaerts, M., & Simons, P. R. J. (1995). *Leren en instructie: Psychologie van de leerling en het leerproces [Learning and Instruction: The psychologie of the learner and the learning process]* (2nd ed.). Assen, The Netherlands: van Gorcum.

Booth, J., MacWhinney, B., Thulborn, K., Sacco, K., Voyvodic, J., & Feldman, H. (2000). Developmental and lesion effects in brain activation during sentence comprehension and mental rotation. *Developmental Neuropsychology, 18*, 139–169.

Bosma, H., Boxtel, M. P. J. v., Ponds, R. W. H. M., Jelicic, M., Houx, P. J., Metsemakers, J., et al. (2002). Engaged lifestyle and cognitive function in middle and old-aged, non-demented persons: a reciprocal association? *Zeitschrift für Gerontologie und Geriatrie, 35*, 575–581.

Boswinkel, N., Baltussen, M., Hoogendijk, W., & Moerlands, F. (2003). *Tussenrapportage project speciaal rekenen: maart 2002 – november 2003*. Utrecht, Den Bosch, Rotterdam: Instituut Freudenthal, Groep KPC, Groep CED.

Boud, D., Keogh, R., & Walker, D. (Eds.). (1985). *Reflection: Turning experience into learning*. London: Kogan Page.

Bowden, E. M., & Beeman, M. J. (1998). Getting the right idea: Semantic activation in the right hemisphere may help solve insight problems. *Psychological Science, 9*, 435–440.

Bowden, E. M., Jung-Beeman, M., Fleck, J., & Kounios, J. (2005). New approaches to demystifying insight. *Trends in Cognitive Sciences, 9*, 322–328.

Bransford, J. D., Brown, A. L., & Cocking, R. R. (Eds.). (1999). *How people learn: Brain, mind, experience, and school*. Washington, DC: National Academy Press.

Bright, P., Moss, H., & Tyler, L. K. (2004). Unitary vs multiple semantics: PET studies of word and picture processing. *Brain and Language, 89*, 417–432.

Buccino, G., Vogt, S., Ritzl, A., Fink, G. R., Zilles, K., Freund, H.-J., et al. (2004). Neural circuits underlying imitation learning of hand actions: An event-related fMRI study. *Neuron, 42*, 323–334.

Bunge, S. A., Klingberg, T., Jacobsen, R. B., & Gabrieli, J. D. E. (2000). A resource model of the neural basis of executive working memory. *Proceeding of the National Academy of Sciences of the United States of America, 97*, 3573–3578.

Buonomano, D. V., & Merzenich, M. M. (1998). Cortical plasticity: from synapses to maps. *Annual Reviews of Neurosciences, 21*, 149–186.

Butler, D., & Winne, P. (1995). Feedback and self-regulated learning: A theoretical synthesis. *Review of Educational Research, 65*, 245–281.

Butterworth, B. (2005). The development of arithmetical abilities. *Journal of Child Psychology and Psychiatry, 46*, 3–18.

Byrnes, J., & Fox, N. (1998). The educational relevance of research in cognitive neuroscience. *Educational Psychology Review, 10*, 297–342.

Caine, R. N., & Caine, G. (1991). *Making connections: Teaching and the human brain.* Alexandria, VA: Association for Supervision and Curriculum Development.

Calfee, R., & Norman, K. (1998). Psychological perspectives on the early reading wars: The case of phonological awareness. *Teachers College Record, 100*, 242–274.

Carey, S. (2001). Cognitive foundations of arithmetic: Evolution and ontogenesis. *Mind & Language, 16*, 37–55.

Carter, C. S., Braver, T. S., Barch, D. M., Botvinick, M. M., Noll, D., & Cohen, J. D. (1998). Anterior cingulate cortex, error detection, and the on-line monitoring of performance. *Science, 280*, 747–749.

Carter, C. S., Mintun, M., & Cohen, J. D. (1995). Inference and facilitation effects during selective attention: An $H_2{}^{15}O$ PET study of Stroop task performance. *Neuroimage, 2*, 264–272.

Casey, B. J., Getz, S., & Galvan, A. (2008). The adolescent brain. *Developmental Review, 28*, 62–77.

Castelli, F., Glaser, D. E., & Butterworth, B. (2006). Discrete and analogue quantity processing in the parietal lobe: A functional MRI study. *Proceedings of the National Academy of Sciences of the United States of America, 103*, 4693–4698.

Chinn, C. A., & Brewer, W. F. (1993). The role of anomalous data in knowledge acquisition: A theoretical framework and implications for science instruction. *Review of Educational Research, 63*, 1–51.

Church, J. A., Coalson, R. S., Lugar, H. M., Petersen, S. E., & Schlaggar, B. L. (2008). A developmental fMRI study of reading and repetition reveals changes in phonological and visual mechanisms over age. *Cerebral Cortex, 18*, 2054–2065.

Cicerone, K. D., Dahlberg, C., Malec, J. F., Langenbahn, D. M., Felicetti, T., Kneipp, S., et al. (2005). Evidence-based cognitive rehabilitation: updated review of the literature from 1998 through 2002. *Archives of Physical Medicine and Rehabilitation, 86*, 1681–1692.

Collette, F., Hogge, M., Salmon, E., & van der Linden, M. (2006). Exploration of the neural substrates of executive functioning by functional neuroimaging. *Neuroscience, 139*, 209–221.

Collins, A., Brown, J. S., & Newman, S. E. (1989). Cognitive apprenticeship: Teaching the crafts of reading, writing, and mathematics. In L. B. Resnick (Ed.), *Knowing, learning, and instruction* (pp. 453–494). Hillsdale, NL: Erlbaum.

Coltheart, M., Curtis, B., Atkins, P., & Haller, M. (1993). Models of reading aloud: Dual-route and parallel-distributed-processing approaches. *Psychological Review, 100*, 589–608.

Coltheart, M., Masterson, J., Byng, S., Prior, M., & Riddoch, J. (1983). Surface dyslexia. *Quarterly journal of experimental psychology, The A, Human experimental psychology, 35*, 469–495.

Coltheart, M., & Rastle, K. (1994). Serial processing in reading aloud: Evidence for dual-route models of reading. *Journal of Experimental Psychology: Human Perception and Performance, 20*, 1197–1211.

Coltheart, M., Rastle, M., Perry, C., Langdon, R., & Zeigler, J. C. (2001). DRC: A dual route cascaded model of visual word recognition and reading aloud. *Psychological Review, 108*, 201–256.

Cowan, N. (2001). The magical number 4 in short-term memory: A reconsideration of mental storage capacity. *Behavioral and Brain Sciences, 24*, 87–114.

Craighero, L., Bello, A., Fadiga, L., & Rizzolatti, G. (2002). Hand action preparation influences the responses to hand pictures. *Neuropsychologia, 40*, 492–502.

Crone, E. A., & Huizinga, M. (2006). Advances in developmental cognitive neuroscience [special issue]. *Neuropsychologia, 44*.

Crottaz-Herbette, S., Anagnoson, R. T., & Menon, V. (2004). Modality effects in verbal working memory: Differential prefrontal and parietal responses to auditory and visual stimuli. *NeuroImage, 21*, 340–351.

D'Esposito, M., Postle, B. R., Jonides, J., & Smith, E. E. (1999). The neural substrate and temporal dynamics of interference effects in working memory as revealed by event-related functional

MRI. *Proceedings of the National Academy of Sciences of the United States of America, 96*, 7514–7519.

de Jong, T. (2006). Computer simulations: Technological advances in inquiry learning. *Science, 312*, 532–533.

de Jong, T., Ainsworth, S., Dobson, M., van der Hulst, A., Levonen, J., Reimann, P., et al. (1998). Acquiring knowledge in science and mathematics: the use of multiple representations in technology based learning environments. In M. v. Someren, P. Reimann, H. Boshuizen, & T. de Jong (Eds.), *Learning with multiple representations* (pp. 9–41). Oxford: Elsevier Science.

Dehaene, S. (1996). The organization of brain activations in number comparison: Event-related potentials and the additive-factors method. *Journal of Cognitive Neuroscience, 8*, 47–68.

Dehaene, S., Dupoux, E., & Mehler, J. (1990). Is numerical comparison digital? Analogical and symbolic effects in two-digit number comparison. *Journal of Experimental Psychology: Human Perception and Performance, 16*, 626–641.

Dehaene, S., Piazza, M., Pinel, P., & Cohen, L. (2003). Three parietal circuits for number processing. *Cognitive Neuropsychology, 20*, 487–506.

Delazer, M., Domahs, F., Bartha, L., Brenneis, C., Lochy, A., Trieb, T., et al. (2003). Learning complex arithmetic – an fMRI study. *Cognitive Brain Research, 18*, 76–88.

Demb, J. B., Boynton, G. M., & Heeger, D. J. (1997). Brain activity in visual cortex predicts individual differences in reading performance. *Proceedings of the National Academy of Sciences of the United States of America, 94*, 13363–13366.

Destrebecqz, A., & Cleeremans, A. (2001). Can sequence learning be implicit? New evidence with the process dissociation procedure. *Psychonomic Bulletin & Review, 8*, 343–350.

Destrebecqz, A., Peigneux, P., Laureys, S., Degueldre, C., Del Fiore, G., Aerts, J., et al. (2005). The neural correlates of implicit and explicit sequence learning: Interacting networks revealed by the process dissociation procedure. *Learning & Memory, 12*, 480–490.

Dolcos, F., & Cabeza, R. (2002). Event-related potentials of emotional memory: Encoding pleasant, unpleasant, and neutral pictures. *Cognitive, Affective, & Behavioral Neuroscience, 2*, 252–263.

Dosenbach, N. U. F., Fair, D. A., Miezin, F. M., Cohen, A. L., Wenger, K. K., Dosenbach, R. A. T., et al. (2007). Distinct brain networks for adaptive and stable task control in humans. *Proceedings of the National Academy of Sciences of the United States of America, 104*, 11073–11078.

Draganski, B., Gaser, C., Busch, V., Schuierer, G., Bogdahn, U., & May, A. (2004). Changes in grey matter induced by training. *Nature, 427*, 311–312.

Duffy, T. M., & Jonassen, D. H. (1992). Constructivism: New implications for instructional technology. In T. M. Duffy, & D. H. Jonassen (Eds.), *Constructivism and the technology of instruction; A conversation* (pp. 7–12). Hillsdale, New Jersey: Lawrence Erlbaum Associates.

Eden, G. F., VanMeter, J. W., Rumsey, J. M., Maisog, J. M., Woods, R. P., & Zeffiro, T. A. (1996). Abnormal processing of visual motion in dyslexia revealed by functional brain imaging. *Nature, 382*, 66–69.

Ehri, L., Nunes, S. R., Stahl, S. A., & Willows, D. M. (2001). Systematic phonics instruction helps students learn to read: Evidence from the national reading panel's meta-analysis. *Review of Educational Research, 71*, 393–447.

Ertmer, P. A., & Newby, T. J. (1996). The expert learner: strategic, self-regulated, and reflective. *Instructional Science, 24*, 1–24.

Feigenson, L., Dehaene, S., & Spelke, E. (2004). Core systems of number. *Trends in Cognitive Sciences, 8*, 307–314.

Ferdinand, N. K., Mecklinger, A., & Kray, J. (2008). Error and deviance processing in implicit and explicit sequence learning. *Journal of Cognitive Neuroscience, 20*, 629–642.

Fernandez-Duque, D., Baird, J. A., & Posner, M. I. (2000). Executive attention and metacognitive regulation. *Consciousness and Cognition, 9*, 288–307.

Fischer, S., Wilhelm, I., & Born, J. (2007). Developmental differences in sleep's role for implicit off-line learning: Comparing children with adults. *Journal of Cognitive Neuroscience, 19*, 214–227.

Flavell, J. H. (1971). First discussant's comments: What is memory development the development of? *Human Development, 14*, 272–278.

Fliessbach, K., Weis, S., Klaver, P., Elger, C. E., & Weber, B. (2006). The effect of word concreteness on recognition memory. *Neuroimage, 32*, 1413–1421.

Fox, E. (2002). Processing emotional facial expressions: The role of anxiety and awareness. *Cognitive, Affective, & Behavioral Neuroscience, 2*, 52–63.

Frost, R. (1998). Toward a strong phonological theory of visual word recognition: True issues and false trails. *Psychological Bulletin, 123*, 71–99.

Frost, R., Katz, L., & Bentin, S. (1987). strategies for visual word recognition and orthographical depth: a multilingual comparison. *Journal of Experimental Psychology: Human Perception and Performance, 13*, 104–115.

Fugelsang, J. A., & Dunbar, K. N. (2005). Brain-based mechanisms underlying complex causal thinking. *Neuropsychologica, 43*, 1204–1213.

Funnell, E. (1983). Phonological processes in reading: New evidence from acquired dyslexia. *British Journal of Psychology, 74*, 159–180.

Gaillard, W., Hertz-Pannier, L., Mott, S., Barnett, A., LeBihan, D., & Theodore, W. (2000). Functional anatomy of cognitive development fMRI of verbal fluency in children and adults. *Neurology, 54*, 180–185.

Galaburda, A. M., Menard, M. T., & Rosen, G. D. (1994). Evidence for aberrant auditory anatomy in developmental dyslexia. *Proceedings of the National Academy of Sciences of the United States of America, 91*, 8010–8013.

Galaburda, A. M., Sherman, G. F., Rosen, G. D., Aboitiz, F., & Geschwind, N. (1985). Developmental dyslexia: four consecutive patients with cortical anomalies. *Annals of Neurology, 18*, 222–233.

Gazzola, V., Rizzolatti, G., Wicker, B., & Keysers, C. (2007). The anthropomorphic brain: The mirror neuron system responds to human and robotic actions. *NeuroImage, 35*, 1674–1684.

Geary, D. (2007). Educating the evolved mind: Conceptual foundations for an evolutionary educational psychology. In J. S. Carlson, & J. R. Levin (Eds.), *Psychological perspectives on contemporary educational issues* (pp. 1–99). Greenwich, CT: Information Age Publishing.

Geary, D. C. (1994). Children's mathematical development: Research and practical applications.

Geary, D. C., Hamson, C. O., & Hoard, M. K. (2000). Numerical and arithmetical cognition: A longitudinal study of process and concept deficits in children with learning disability. *Journal of Experimental Child Psychology, 77*, 236–263.

Geary, D. C., & Hoard, M. K. (2001). Numerical and arithmetical deficits in learning-disabled children: Relation to dyscalculia and dyslexia. *Aphasiology, 15*, 635–647.

Genesee, F., Nicoladis, E., & Paradis, J. (1995). Language differentiation in early bilingual development. *Journal of Child Language, 22*, 611–631.

Giedd, J. N. (2004). Structural magnetic resonance imaging of the adolescent brain. *Annals of the New York Academy of Sciences*, 77–85.

Giedd, J. N. (2008). The teen brain: insights from neuroimaging. *Journal of Adolescent Health, 42*, 335–343.

Gogtay, N., Giedd, J. N., Lusk, L., Hayashi, K. M., Greenstein, D., Vaituzis, A. C., et al. (2004). Dynamic mapping of human cortical development during childhood through early adulthood. *Proceedings of the National Academy of Sciences of the United States of America, 101*, 8174–8179.

Goswami, U. (2004). Neuroscience and education. *British Journal of Educational Psychology, 74*, 1–14.

Goswami, U. (2006). Neuroscience and education: from research to practice? *Nature Reviews Neuroscience, 7*, 406–411.

Goswami, U., Thomson, J., Richardson, U., Stainthorp, R., Hughes, D., & Rosen, S. (2002). Amplitude envelope onsets and developmental dyslexia: A new hypothesis. *Proceedings of the National Academy of Sciences of the United States of America, 99*, 10911–10916.

Grabner, R. H., Neubauer, A. C., & Stern, E. (2006). Superior performance and neural efficiency: The impact of intelligence and expertise. *Brain Research Bulletin, 69*, 422–439.

Graddol, D. (2004). The future of language. *Science, 303*, 1329–1331.

Guttorm, T. K., Leppanen, P. H., Poikkeus, A. M., Eklund, K. M., Lyytinen, P., & Lyytinen, H. (2005). Brain event-related potentials (ERP's) measured at birth predict later language development in children with and without familial risk for dyslexia. *Cortex, 41*, 291–303.

Habib, M. (2000). The neurological basis of developmental dyslexia. An overview and working hypothesis. *Brain*, 123, 2373–2399.

Henson, R. (2006). Forward inference using functional neuroimaging: Dissociations versus associations. *TRENDS in Cognitive Sciences, 10*, 64–69.

Höffler, T. N., & Leutner, D. (2007). Instructional animation versus static pictures: A meta-analysis. *Learning and Instruction, 17*, 722–738.

Hurley, S. (2008). The shared circuits model (SCM): How control, mirroring, and simulation can enable imitation, deliberation, and mind reading. *Behavioral and Brain Sciences, 31*, 1–58.

Iacoboni, M., Woods, R., Brass, M., Bekkering, H., Mazziotta, J., & Rizzolatti, G. (1999). Cortical mechanisms of human imitation. *Science, 286*, 2526–2528.

Immordino-Yang, M. H., & Damasio, A. (2007). We feel, therefore we learn: The relevance of affective and social neuroscience to education. *Mind, Brain, and Education, 1*, 3–10.

Ischebeck, A., Zamarian, L., Siedentopf, C., Koppelstatter, F., Benke, T., Felber, S., et al. (2006). How specifically do we learn? Imaging the learning of multiplication and subtraction. *NeuroImage, 30*, 1365–1375.

Jansma, J. M., Ramsey, N. F., Slagter, H. A., & Kahn, R. S. (2001). Functional anatomical correlates of controlled and automatic processing. *Journal of Cognitive Neuroscience, 13*, 730–743.

Jausovec, N., & Bakracevic, K. (1995). What can heart rate tell us about the creative process? *Creativity Research Journal, 8*, 11–24.

Jeynes, W. H., & Littell, S. W. (2000). A meta-analysis of studies examining the effect of whole language instruction on the literacy of low-SES students. *The Elementary School Journal, 101*, 21–33.

Jobard, G., Crivello, F., & Tzourio-Mazoyer, N. (2003). Evaluation of the dual route theory or reading: A metanalysis of 35 neuroimaging studies. *Neuroimage, 20*, 693–712.

Jolles, J. (2007a). Neurocognitieve ontwikkeling en adolescentie: enkele implicaties voor het onderwijs. *Onderwijsinnovatie*, 30–32.

Jolles, J. (2007b). Over brein & leren: kwaliteit leeromgeving bepalend voor leerproces. *Bij de Les, februari*, 43–45.

Jolles, J., de Groot, R., van Benthem, J., Dekkers, H., de Glopper, C., Uijlings, H., et al. (2005). *Leer het brein kennen (Learning to know the brain)*. The Hague (NL): Netherlands Organization for Scientific Research.

Jolles, J., de Groot, R., van Benthem, J., Dekkers, H., de Glopper, C., Uijlings, H., et al. (2006). *Brain Lessons; A contribution to the international debate on Brain, Learning & Education*. Maastricht: NeuroPsych.

Jonassen, D. H. (1999). Designing constructivist learning environments. In C. M. Reigeluth (Ed.), *Instructional-design theories and models: A new paradigm of instructional theory* (Vol. 2, pp. 215–239). London: Lawrence Erlbaum Associates.

Jonides, J., Schumacher, E. H., Smith, E. E., Lauber, E. J., Awh, E., Minoshima, S., et al. (1997). Verbal working memory load affects regional brain activation as measured by PET. *Journal of Cognitive Neuroscience, 9*, 462.

Jung-Beeman, M., Bowden, E. M., Haberman, J., Frymiare, J. L., Arambel-Liu, S., Greenblatt, R., et al. (2004). Neural activity when people solve verbal problems with insight. *PLoS Biology, 2*, 500–510.

Karp, A., Paillard-Borg, S., Wang, H., Silverstein, M., Winblad, B., & Fratiglioni, L. (2006). Mental, physical and social components in leisure activities equally contribute to decrease dementia risk. *Dementia and Geriatric Cognitive Disorders, 21*, 65–73.

Katz, L., & Frost, R. (1992). The reading process is different for different orthographies: The orthographic depth hypothesis. In R. Frost, & L. Katz (Eds.), *Orthography, phonology, morphology, and meaning: Advances in psychology* (Vol. 94, pp. 67–84). Oxford: North Holland.

Katzir, T., & Paré-Blagoev, J. (2006). Applying cognitive neuroscience research to education: The case of literacy. *Educational Psychologist, 41*, 53–74.

Kennedy, D., & Norman, C. (2005). What don't we know? *Science, 309*, 75–75.

Keysers, C., & Gazzola, V. (2007). Integrating simulation and theory of mind: From self to social cognition. *TRENDS in Cognitive Sciences, 11*, 194–196.

Kim, K. H., Relkin, N. R., Lee, K., & Hirsch, J. (1997). Distinct cortical areas associated with native and second languages. *Nature neuroscience, 388*, 171–174.

Kirchhoff, B. A., & Buckner, R. L. (2006). Functional-anatomic correlates of individual differences in memory. *Neuron, 51*, 263–274.

Kirk, E. P., & Ashcraft, M. H. (2001). Telling stories: The perils and promise of using verbal reports to study math strategies. *Journal of Experimental Psychology: Learning, Memory, and Cognition, 27*, 157–175.

Kolb, B., & Whishaw, I. Q. (2008). *Fundamentals of human neuropsychology* (6th ed.). New York: Worth Publishers.

Könings, K., Brand-Gruwel, S., van Merriënboer, J. J. G., & Broers, N. (2008). Does a new learning environment come up to students' expectations? A longitudinal study. *Journal of Educational Psychology, 100*, 535–548.

Koriat, A. (2000). The Feeling of Knowing: Some metatheoretical implications for consciousness and control. *Consciousness and Cognition, 9*, 149–171.

Kovelman, I., & Petitto, L. A. (2002). *Bilingual babies' maturational and linguistic milestones as a function of their age of first exposure to two languages.* Paper presented at the Conference for the Society of Neuroscience, Orlango, FL.

Kroesbergen, E. H., & van Luit, J. E. H. (2005). Constructivist mathematics education for students with mild mental retardation. *European Journal of Special Needs Education, 20*, 107–116.

Lang, S., Kanngieser, N., Jaskowski, P., Haider, H., Rose, M., & Verleger, R. (2006). Precursors of insight in event-related brain potentials. *Journal of Cognitive Neuroscience, 18*, 2152–2166.

Lerner, J., & Kline, F. (2006). *Learning disabilities and related disorders: characteristics and teaching strategies* (10th ed.). Boston: Houghton Miffin.

Lipton, J. (2005). Origins of number sense: The development of numerical cognition in infants and preschoolers. *Dissertation Abstracts International: Section B: The Sciences and Engineering, 66.*

Livingstone, M. S., Rosen, G. D., Drislane, F. W., & Galaburda, A. M. (1991). Physiological and anatomical evidence for a magnocellular defect in developmental dyslexia. *Proceedings of the National Academy of Sciences of the United States of America, 88*, 7943–7947.

Locke, H. S., & Braver, T. S. (2008). Motivational influences on cognitive control: Behavior, brain activation, and individual differences. *Cognitive, Affective, & Behavioral Neuroscience, 8*, 99–112.

Lovegrove, W. J., Bowling, A., Badcock, D., & Blackwood, M. (1980). Specific reading disability: differences in contrast sensitivity as a function of spatial frequency. *Science, 210*, 439–440.

Low, R., & Sweller, J. (2005). The modality principle in multimedia learning. In R. E. Mayer (Ed.), *The Cambridge Handbook of Multimedia Learning*. New York: Cambridge University Press.

Maehr, M. L. (2001). Goal theory is not dead – Not yet, anyway: A reflection on the special issue. *Educational Psychology Review, 13*, 177–185.

Maguire, E. A., Gadian, D. G., Johnsrude, I. S., Good, C. D., Ashburner, J., Frackowiak, R., S. J., et al. (2000). Navigation-related structural change in the hippocampi of taxi drivers. *PNAS, 97*, 4398–4403.

Maier, N. R. F. (1931). Reasoning in humans II: The solution of a problem and its appearance in consciousness. *Journal of Comparative Psychology, 12*, 181–194.

Manis, F. R., Seidenberg, M. S., Doi, L. M., McBride-Chang, C., & Petersen, A. (1996). On the bases of two subtypes of developmental [corrected] dyslexia. *Cognition, 58*, 157–195.

Manlove, S., Lazonder, A. W., & de Jong, T. (2007). Software scaffolds to promote regulation during scientific inquiry learning. *Metacognition and Learning*, *2*, 141–155.
Marshall, J.-C., & Newcombe, F. (1981). Lexical access: A perspective from pathology. *Cognition*, *10*, 209–214.
Marshall, L., Helgadottir, H., Molle, M., & Born, J. (2006). Boosting slow oscillations during sleep potentiates memory. *Nature*, *444*, 610–613.
Marshall, U., & Born, J. (2007). The contribution of sleep to hippocampus-dependent memory consolidation. *Trends in Cognitive Sciences*, *11*, 442–450.
Maurer, U., Brem, S., Bucher, K., Kranz, F., Benz, R., Steinhausen, H. C., et al. (2007). Impaired tuning of a fast occipito-temporal response for print in dyslexic children learning to read. *Brain*, *130*, 3200–3210.
Mayer, R. E. (2001). *Multimedia learning*. New York: Cambridge University Press.
Mazzocco, M. M. M., & Kover, S. T. (2007). A longitudinal assessment of executive function skills and their association with math performance. *Child Neuropsychology*, *13*, 18–45.
McLaughlin, J., Osterhout, L., & Kim, A. (2004). Neural correlates of second-language word learning: minimal instruction produces rapid change. *Nature neuroscience*, *7*, 703–704.
Meltzoff, A. N., & Prinz, W. (2002). *The imitative mind: Development, evolution, and brain bases*. Cambridge, UK: Cambridge University Press.
Merkx, F., & van Koten, R. C. (in press). *The development of transdisciplinary learning science. Where do we stand*. The Hague: Rathenau Institute.
Merzenich, M. M., Jenkins, W. M., Johnston, P., Schrenier, C., Miller, S. L., & Tallal, P. (1996). Temporal processing deficits of language-learning impaired children ameliorated by training. *Science*, *271*, 77–81.
Metcalfe, J. (1986). Feeling of Knowing in memory and problem solving. *Journal of Experimental Psychology: Learning, Memory, and Cognition*, *12*, 288–294.
Metcalfe, J., & Wiebe, D. (1987). Intuition in insight and non-insight problems.*Memory and Cognition*, *15*, 238–246.
Mezzacappa, E. (2004). Alterting, orienting, and executive attention: Developmental properties and sociodemographic correlates in an epidemiological sample of young urban children. *Child Development*, *75*, 1373–1386.
Miller, G. A. (1956). The magical number seven, plus or minus two: Some limits on our capacity to process information. *Psychological Review*, *63*, 81–97.
Molfese, D.-L. (2000). Predicting dyslexia at 8 years of age using neonatal brain responses. *Brain and Language*, *72*, 238–245.
Nakada, T., Fujii, Y., & Kwee, I. L. (2001). Brain Strategies for Reading in the Second Language are Determined by the First Language. *Neuroscience Research*, *40*, 351–358.
Nieuwenhuis, S., Ridderinkhof, K. R., Blom, J., Band, G. P. H., & Kok, A. (2001). Error-related brain potentials are differentially related to awareness of response errors: Evidence from an anti-saccade task. *Psychophysiology*, *38*, 752–760.
Noble, K. G., Norman, F., & Farah, M. J. (2005). Neurocognitive correlates of socioeconomic status in kindergarten children. *Developmental Science*, *8*, 74–87.
Noble, K. G., Tottenham, N., & Casey, B. J. (2005). Neuroscience Perspective on Disparities in School Readiness and Cognitive Achievement. *The Future of Children*, *15*, 71–89.
Norris, C. J., Chen, E. E., Zhu, D. C., Small, S. L., & Cacioppo, J. T. (2004). The Interaction of Social and Emotional Processes in the Brain. *Journal of Cognitive Neuroscience*, *16*, 1818–1829.
OECD. (2002). *Understanding the brain: towards a new learning science*. Paris: Organisation for Economic Co-operation and Development.
OECD. (2007). *Understanding the Brain: The Birth of a Learning Science*. Paris: Office of Economic Cooperation and Development.
Ohlsson, S. (1992). Information processing explanations of insight and related phenomena. In M. T. Keane, & K. J. Gilhooly (Eds.), *Advances in the psychology of thinking*(pp. 1–43). London: Harvester-Wheatsheaf.
Paivio, A. (1979). *Imagery and verbal processes*. Hillsdale, NJ: Erlbaum.

Paivio, A. (1986). *Mental representations*. New York: Oxford University Press.

Paivio, A. (1991). Dual Coding Theory: Retrospect and current status. *Canadian Journal of Psychology, 45*, 255–287.

Palmer, S. E. (1978). Fundamental aspects of cognitive representation. In E. Rosch, & B. B. Lloyd (Eds.), *Cognition and categorization* (pp. 259–303). Hillsdale, NJ: LEA.

Paris, S. (2005). Reinterpreting the development of reading skills. *Reading Research Quarterly, 40*, 184–202.

Paulesu, E., Demonet, J., Fazio, F., McCrory, E., Chanoine, V., Brunswick, N., et al. (2001). Dyslexia: Cultural Diversity and Biological Unity. *Science, 291*, 2165–2167.

Paulesu, E., Frith, U., Snowling, M., Gallagher, A., Frackowiak, R. S., & Frith, C. D. (1996). Is developmental dyslexia a disconnection syndrome? Evidence from PET scanning *Brain, 119*, 143–157.

Paulesu, E., McCrory, E., Fazio, F., Menoncello, L., Brunswick, N., Cappa, S. F., et al. (2000). A cultural effect on brain function. *Nature Neuroscience, 3*, 91–96.

Paus, L. (2005). Mapping brain maturation and cognitive development during adolescence. *Trends in Cognitive Sciences*, 60–68.

Pekrun, R. (2005). Progress and open problems in educational emotion research. *Learning and Instruction, 15*, 497–506.

Pekrun, R., Goetz, T., Titz, W., & Perry, R. P. (2002). Academic emotions in students' self-regulated learning and achievement: a program of quantitative and qualitative research. *Educational Psychologist, 37*, 91–106.

Pennington, B. F. (1999). Toward an integrated understanding of dyslexia: Genetic, neurological, and cognitive mechanisms. *Development and Psychopathology, 11*, 629–654.

Peters, K. R., Smith, V., & Smith, C. T. (2007). Changes in sleep architecture following motor learning depend on initial skill level. *Journal of Cognitive Neuroscience, 19*, 817–829.

Peterson, S. E., & Schreiber, J. B. (2006). An attributional analysis of personal and interpersonal motivation for collaborative projects. *Journal of Educational Psychology, 98*, 777–787.

Petitto, L. A., Baker, S., Baird, J. A., Kovelman, I., & Norton, E. (2004). *Near-infrared Spectroscopy studies of children and adults during language processing.* Paper presented at the International Workshop on Near-Infrared Spectroscopy, Cambridge, MA.

Petitto, L. A., & Dunbar, K. N. (in press). New findings from educational neuroscience on bilingual brains, scientific brains, and the educated mind. In K. Fischer, & T. Katz (Eds.), *Building Usable Knowledge in Mind, Brain, and Education*. Cambridge: Cambridge University Press.

Petitto, L. A., Kovelman, I., & Harasymowycz, U. (2003). *Bilingual language development: Does learning the new damage the old?* Paper presented at the Biannual meeting of the Society for Research in Child Development, Tampa, FL.

Poldrack, R. A. (2006). Can cognitive processes be inferred from neuroimaging data? *TRENDS in Cognitive Sciences, 10*, 59–63.

Posner, M. I., & Rothbart, M. K. (2005). Influencing brain networks: implications for education. *Trends in Cognitive Sciences, 9*.

Rae, C., Harasty, J. A., Dzendrowskyj, T. E., Talcott J. B., Simpson, J. M., Blamire, A. M., et al. (2002). Cerebellar morphology in developmental dyslexia. *Neuropsychologia, 40*, 1285–1292.

Rapp, B. C., Hillis, A. E., & Caramazza, A. (1993). The role of representations in cognitive theory: More on multiple semantics and the agnosias. *Cognitive Neuropsychology, 10*, 235–249.

Reber, A. S. (1989). Implicit learning and tacit knowledge. *Journal of Experimental Psychology-General, 118*, 219–235.

Reber, P. J., Gitelman, D. R., Parrish, T. B., & Mesulam, M. M. (2003). Dissociating explicit and implicit category knowledge with fMRI. *Journal of Cognitive Neuroscience, 15*, 574–583.

Reder, L. M., & Schunn, C. D. (1996). Metacognition does not imply awareness: Strategy choice is governed by implicit learning and memory. In L. M. Reder (Ed.), *Implicit memory and metacognition* (pp. 45–77). Mahwah, NJ: Erlbaum.

Redlinger, W. E., & Park, T. Z. (1980). Language mixing in young bilinguals. *Journal of Child Language, 7*, 337–352.

Reiss, J. P., Campbell, D. W., Leslie, W. D., Paulus, M. P., Stroman, P. W., Polimeni, J. O., et al. (2005). The role of the striatum in implicit learning: a functional magnetic resonance imaging study. *Neuroreport, 16*, 1291–1295.

Renkl, A. (2005). The worked-out examples principle in multimedia learning. In R. E. Mayer (Ed.), *The Cambridge handbook of multimedia learning* (pp. 229–246). New York: Cambridge University Press.

Richardson, U., Thomson, J. M., Scott, S. K., & Goswami, U. (2004). Auditory processing skills and phonological representation in dyslexic children. *Dyslexia, 10*, 215–233.

Rizzolatti, G., & Craighero, L. (2004). The mirror-neuron system. *Annual Review of Neuroscience, 27*, 169–192.

Robertson, E. M., Pascual-Leone, A., & Press, D. Z. (2004). Awareness modifies the skill-learning benefits of sleep. *Current Biology, 14*, 208–212.

Robertson, I. H., Manly, T., Andrade, J., Baddeley, B. T., & Yiend, J. (1997). "Oops!": Performance correlates of everyday attentional failures in traumatic brain injured and normal subjects. *Neuropsychologica, 35*, 747–758.

Robertson, I. H., & Murre, J. M. J. (1999). Rehabilitation of brain damage: brain plasticity and principles of guided recovery. *Psychological Bulletin, 125*, 544–575.

Rose, M., Haider, H., Weiller, C., & Buchel, C. (2004). The relevance of the nature of learned associations for the differentiation of human memory systems. *Learning & Memory, 11*, 145–152.

Rourke, B.-P., & Conway, J.-A. (1997). Disabilities of arithmetic and mathematical reasoning: Perspectives from neurology and neuropsychology. *Journal of Learning Disabilities, 30*, 34–46.

Rousselle, L., & Noel, M.-P. (2007). Basic numerical skills in children with mathematics learning disabilities: A comparison of symbolic vs non-symbolic number magnitude. *Cognition, 102*, 361–395.

Rumsey, J. M., Casanova, M., Mannheim, G. B., Patronas, N., de Vaughn, N., Hamburger, S. D., et al. (1996). Corpus callosum morphology, as measured with MRI, in dyslexic men. *Biological Psychiatry, 39*, 769–775.

Ruijssenaars, A. J. J. M., van Luit, J. E. H., & van Lieshout, E. C. D. M. (2004). *Rekenproblemen en dyscalculie: theorie, onderzoek, diagnostiek en behandeling.* Rotterdam, The Netherlands: Lemniscaat b.v.

Saetrevik, B., Reber, R., & Sannum, P. (2006). The utility of implicit learning in the teaching of rules. *Learning and Instruction, 16*, 363–373.

Sandkuhler, S., & Bhattacharya, J. (2008). Deconstructing insight: EEG correlates of insightful problem solving. *PLoS One*, 1–12.

Schendan, H. E., Searl, M. M., Melrose, R. J., & Stern, C. E. (2003). An fMRI study of the role of the medial temporal lobe in implicit and explicit sequence learning. *Neuron, 37*, 1013–1025.

Schneider, W., & Shiffrin, R. M. (1977). Controlled and automatic human information processing: I. Detection, search and attention. *Psychological Review, 84*, 1–66.

Schnyer, D., Nicholls, L., & Verfaellie, M. (2005). The role of VMPC in metamemorial judgements of content retrievability. *Journal of Cognitive Neuroscience, 17*, 832–846.

Schön, D. (1991). *The reflective practitioner* (1st ed.). Aldershot, Hants England: Arena Ashgate Publishing Limited.

Schraw, G. (1998). Promoting general metacognitive awareness. *Instructional Science, 26*, 113–125.

Seufert, T. (2003). Supporting coherence formation in learning from multiple representations. *Learning and Instruction, 13*, 227–237.

Shalev, R. S., & Gross-Tsur, V. (2001). Developmental dyscalculia. *Pediatric Neurology, 24*, 337–342.

Shallice, T., Warrington, E.-K., & McCarthy, R. (1983). Reading without semantics. *The Quarterly Journal of Experimental Psychology A: Human Experimental Psychology, 35*, 111–138.

Shanks, D. R., & StJohn, M. F. (1994). Characteristics of dissociable human learning-systems. *Behavioral and Brain Sciences, 17*, 367–395.

Shaywitz, B. A., Shaywitz, S. E., Pugh, K. R., Fulbright, R. K., Skudlarski, P., Mencl, W. E., et al. (2001). The functional neural architecture of components of attention in language-processing tasks. *NeuroImage*, *13*, 601–612.

Shiffrin, R. M., & Schneider, W. (1977). Controlled and automatic human information processing: II. Perceptual learning, automatic attending, and a general theory. *Psychological Review*, *84*, 127–190.

Shimamura, A. P. (2000). Towards a cognitive neuroscience of metacognition. *Consciousness and Cognition*, *9*, 313–323.

Simos, P. G., Fletcher, J. M., Bergman, E., Breier, J. I., Foorman, B. R., Castillo, E. M., et al. (2002). Dyslexia-specific brain activation profile becomes normal following successful remedial training. *Neurology*, *58*, 1203–1213.

Simpson, J. R., Öngür, D., Akbudak, E., Conturo, T. E., Ollinger, J. M., Snyder, A. Z., et al. (2000). The emotional modulation of cognitive processing: An fMRI study. *Journal of Cognitive Neuroscience*, *12*, 157–170.

Siok, W. T., Niu, Z. D., Jin, Z., Perfetti, C. A., & Tan, L. H. (2008). A structural-functional basis for dyslexia in the cortex of Chinese readers. *Proceedings of the National Academy of Sciences of the United States of America*, *105*, 5561–5566.

Siok, W. T., Perfetti, C. A., Jin, Z., & Tan, L. H. (2004). Biological abnormality of impaired reading is constrained by culture. *Nature*, *43*, 71–76.

Sperling, A. J., Zhong-LIn, L., Manis, R., & Seidenberg, M. S. (2005). Deficits in perceptual noise exclusion in developmental dyslexia. *Nature Neuroscience*, *7*, 862–863.

Starkey, P., & Cooper, R. G. (1980). Perception of numbers by human infants. *Science*, *210*, 1033–1035.

Stein, J. (2001). The magnocellular theory of developmental dyslexia. *Dyslexia*, *7*, 12–36.

Steinberg, L. (2005). Cognitive and affective development in adolescence. *Trends in Cognitive Science*, *9*, 69–74.

Steinberg, L. (2008). A social neuroscience perspective on adolescent risk-taking. *Developmental Review*, *28*, 78–106.

Stern, E., Grabner, R., & Schumacher, R. (2006). *Educational Research and Neurosciences – Expectations, Evidence, Research Prospects*. Bonn, Berlin: Federal Ministry of Education and Research (BMBF).

Stock, P., Desoete, A., & Roeyers, H. (2006). Focussing on mathematical disabilities: A search for definition, classification and assessment. In S. V. Randall (Ed.), *Learning Disabilities: New Research* (pp. 29–62). Hauppauge NY: Nova Science Publishers, Inc.

Stone, G. O., Vanhoy, M., & van Orden, G. C. (1997). Perception is a two-way street: Feedforward and feedback phonology in visual word recognition. *Journal of Memory and Language*, *36*, 337–359.

Streefland, L. (1986). Rational analysis of realistic mathematics education as a theoretical source of psychology: Fractions as a paradigm. *European Journal of Psychology of Education*, *1*, 67–82.

Sun, R., Merrill, E., & Peterson, T. (2001). From implicit skills to explicit knowledge: a bottom-up model of skill learning. *Cognitive Science*, *25*, 203–244.

Sweller, J. (1988). Cognitive load during problem solving: Effects on learning. *Cognitive Science*, *12*, 257–285.

Sweller, J., & Sweller, S. (2006). Natural information processing systems. *Evolutionary Psychology*, *4*, 434–458.

Sweller, J., van Merriënboer, J. J. G., & Paas, F. (1998). Cognitive architecture and instructional design. *Educational Psychology Review*, *10*, 251–295.

Tai, Y., Scherfler, C., Brooks, D., Sawamoto, N., & Castiello, U. (2004). The human premotor cortex is "mirror" only for biological actions. *Current Biology*, *14*, 117–120.

Talcott, J. B., Hansen, P. C., Assoku, E. L., & Stein, J. F. (2000). Visual motion sensitivity in dyslexia: evidence for temporal and energy integration deficits. *Neuropsychologia*, *38*, 935–943.

Tallal, P. (1980). Auditory temporal perception, phonics, and reading disabilities in children. *Brain and Language, 9*, 182–198.

Tallon-Baudry, C., & Bertrand, O. (1999). Oscillatory gamma activity in humans and its role in object representation. *Trends in Cognitive Sciences, 3*, 151–162.

Tan, L. H., Spinks, J. A., Feng, C. M., Siok, W. T., Perfetti, C. A., & Xiong, J. (2003). Neural systems of second language reading are shaped by native language. *Neuropsychologia, 18*, 158–166.

Taub, E. (2004). Harnessing brain plasticity through behavioural techniques to produce new treatments in neurorehabilitation. *American Psychologist, 59*, 692–704.

Taub, E., Uswatte, G., King, D. K., Morris, D., Crago, J. E., & Chatterjee, A. (2006). A placebo-controlled trial of constraint-induced movement therapy for upper extremity after stroke. *Stroke, 37*, 1045–1049.

Tettamanti, M., Buccino, G., Saccuman, M. C., Gallese, V., Danna, M., Scifo, P., et al. (2005). Listening to action-related sentences activates fronto-parietal motor circuits. *Journal of Cognitive Neuroscience, 17*, 273–281.

Thiede, K. W., Anderson, M. C. M., & Therriault, D. (2003). Accuracy of metacognitive monitoring affects learning of texts. *Journal of Educational Psychology, 95*, 66–73.

Thierry, G., & Wu, Y. J. (2007). Brain potentials reveal unconscious translation during foreign-language comprehension. *Proceedings of the National Academy of Sciences of the United States of America, 104*, 12530–12535.

Thomas, K. M., Hunt, R. H., Vizueta, N., Sommer, T., Durston, S., Yang, Y., et al. (2004). Evidence of developmental differences in implicit sequence learning: An fMRI study of children and adults. *Journal of Cognitive Neuroscience, 6*, 1339–1351.

Tokowicz, N., & MacWhinney, B. (2005). Implicit and explicit measures of sensitivity to violations in second language grammar. An event-related potential investigation. *Studies in Second Language Acquisition, 27*, 173–204.

Treffers, A. (1993). Wiskobas and Freudenthal: Realistic mathematics education. *Educational Studies in Mathematics, 25*, 89–108.

Tversky, B., Morrison, J. B., & Betrancourt, M. (2002). Animation: Can it facilitate? *International Journal of Human-Computer Studies, 57*, 247–262.

van den Boom, G., Paas, F., van Merriënboer, J. J. G., & van Gog, T. (2004). Reflection prompts and tutor feedback in a web-based learning environment: Effects on students self-regulated learning competence. *Computers in Human Behavior, 20*, 551–567.

van der Meij, J., & de Jong, T. (2006). Supporting students' learning with multiple representations in a dynamic simulation-based learning environment. *Learning and Instruction, 16*, 199–212.

van Gog, T., Paas, F., Marcus, N., Ayres, P., & Sweller, J. (2009). The mirror-neuron system and observational learning: Implications for the effectiveness of dynamic visualizations. *Educational Psychology Review*.

van Gog, T., Paas, F., & van Merriënboer, J. J. G. (2005). Uncovering expertise-related differences in troubleshooting performance: Combining eye movement and concurrent verbal protocol data. *Applied Cognitive Psychology, 19*, 205–221.

van Gog, T., Paas, F., van Merriënboer, J. J. G., & Witte, P. (2005). Uncovering the problem-solving process: Cued retrospective reporting versus concurrent and retrospective reporting. *Journal of Experimental Psychology: Applied, 11*, 237–244.

van Hell, J. G., & de Groot, A. M. B. (1998). Disentangling context availability and concreteness in lexical decision and word translation. *Quarterly Journal of Experimental Psychology Section a-Human Experimental Psychology, 51*, 41–63.

van Hell, J. G., & Tokowicz, N. (in press). Event-related potentials and second language research: Syntactic processing in late L2 learners at different L2 fluency levels. *Second Language Research*.

van Keulen, M. T. E., & Rispens, J. (in press). Personalised Learning: research directions for neuroscience, cognitive science and educational practice. In M. T. E. van Keulen (Ed.), *Brain*

Visions: future applications of the brain sciences in food, education, interface design and law. The Hague (the Netherlands): Stichting Toekomstbeeld der Techniek.

van Leeuwen, T., Been, P., van Herten, M., Zwarts, F., Maassen, B., & van der Leij, A. (2008). Two-month-old infants at risk for dyslexia do not discriminate /bAk/ from /dAk/: A brain-mapping study. *Journal of Neurolinguistics, 21*, 333–348.

van Merriënboer, J. J. G., & Kirschner, P. A. (2007). *Ten steps to complex learning*. Mahwah, NJ: Erlbaum/Taylor and Francis.

van Merriënboer, J. J. G., & Sweller, J. (2005). Cognitive load theory and instructional design: Recent developments and future directions. *Educational Psychology Review, 17*, 147–177.

Vandenberghe, R., Price, C., Wise, R., Josephs, O., & Frackowiak, R. S. J. (1996). Functional anatomy of a common semantic system for words and pictures. *Nature, 383*, 254–256.

Varma, S., McCandliss, B. D., & Schwartz, D. L. (2008). Scientific and pragmatic challenges for bridging education and neuroscience. *Educational Researcher, 37*, 140–152.

Vogt, S., Taylor, P., & Hopkins, B. (2003). Visuomotor priming by pictures of hand postures: perspective matters. *Neuropsychologia, 41*, 941–951.

Vygotsky, L. S. (1978). *Mind in society*. Cambridge, MA: Harvard University Press.

Wartenburger, I., Heekeren, H. R., Abutalebi, J., Cappa, S. F., Villringer, A., & Perani, D. (2003). Early setting of grammatical processing in the bilingual brain. *Neuron, 37*, 159–170.

Weinberger, N. M. (1995). Dynamic regulation of receptive fields and maps in the adult sensory cortex. *Annual Reviews of Neurosciences, 18*, 129–158.

WHO. (1992). *The ICD-10 classification of mental and behavioural disorders: clinical descriptions and diagnostic guidelines*. Geneva: WHO.

Wilson, R. S., Mendes de Leon, C. F., Barnes, L. L., Schneider, J. A., Bienias, J. L., Evans, D. A., et al. (2002). Participation in cognitively stimulating activities and risk of incident Alzheimer disease. *JAMA, 287*, 742–748.

Wynn, K. (1992). Addition and subtraction by human infants. *Nature, 358*, 749–750.

Xu, F., & Spelke, E. S. (2000). Large number discrimination in 6-month-old infants. *Cognition, 74*, 1–11.

Zeigler, J. C., & Goswami, U. (2006). Becoming literate in different languages: similar problems, different solutions. *Developmental Science, 9*, 429–453.

Ziegler, J. C., Perry, C., Jacobs, A. M., & Braun, M. (2001). Identical words are read differently in different languages. *Psychological Science, 12*, 379–384.

Zimmerman, B. J. (2000). Attaining self-regulation: A social cognitive perspective. In M. Boekaerts, P. Pintrich, & M. Zeidner (Eds.), *Handbook of self regulation* (pp. 13–35). London: Academic Press.

Ziori, E., & Dienes, Z. (2008). How does prior knowledge affect implicit and explicit concept learning? *Quarterly Journal of Experimental Psychology, 61*, 601–624.

Index

GPSR Compliance

The European Union's (EU) General Product Safety Regulation (GPSR) is a set of rules that requires consumer products to be safe and our obligations to ensure this.

If you have any concerns about our products, you can contact us on ProductSafety@springernature.com

In case Publisher is established outside the EU, the EU authorized representative is:

Springer Nature Customer Service Center GmbH
Europaplatz 3
69115 Heidelberg, Germany

Batch number: 10844675

Printed by Printforce, the Netherlands